Lecture Notes in Physics

Founding Editors

Wolf Beiglböck
Jürgen Ehlers
Klaus Hepp
Hans-Arwed Weidenmüller

Volume 1027

Series Editors

Roberta Citro, Salerno, Italy
Peter Hänggi, Augsburg, Germany
Betti Hartmann, London, UK
Morten Hjorth-Jensen, Oslo, Norway
Maciej Lewenstein, Barcelona, Spain
Satya N. Majumdar, Orsay, France
Luciano Rezzolla, Frankfurt am Main, Germany
Angel Rubio, Hamburg, Germany
Wolfgang Schleich, Ulm, Germany
Stefan Theisen, Potsdam, Germany
James D. Wells, Ann Arbor, MI, USA
Gary P. Zank, Huntsville, AL, USA

The series Lecture Notes in Physics (LNP), founded in 1969, reports new developments in physics research and teaching - quickly and informally, but with a high quality and the explicit aim to summarize and communicate current knowledge in an accessible way. Books published in this series are conceived as bridging material between advanced graduate textbooks and the forefront of research and to serve three purposes:

- to be a compact and modern up-to-date source of reference on a well-defined topic;
- to serve as an accessible introduction to the field to postgraduate students and non-specialist researchers from related areas;
- to be a source of advanced teaching material for specialized seminars, courses and schools.

Both monographs and multi-author volumes will be considered for publication. Edited volumes should however consist of a very limited number of contributions only. Proceedings will not be considered for LNP.

Volumes published in LNP are disseminated both in print and in electronic formats, the electronic archive being available at springerlink.com. The series content is indexed, abstracted and referenced by many abstracting and information services, bibliographic networks, subscription agencies, library networks, and consortia.

Proposals should be sent to a member of the Editorial Board, or directly to the responsible editor at Springer:

Dr Lisa Scalone
lisa.scalone@springernature.com

T. R. Govindarajan • Pichai Ramadevi

Geometry and Topology of Low Dimensional Systems

Chern-Simons Theory with Applications

T. R. Govindarajan
Theoretical Physics
Institute of Mathematical Sciences
Chennai, Tamil Nadu, India

School of Interwoven Arts and Sciences
(SIAS), Physics
Krea University
Sri City, Andhra Pradesh, India

Pichai Ramadevi
Department of Physics
Indian Institute of Technology Bombay
Mumbai, India

ISSN 0075-8450 ISSN 1616-6361 (electronic)
Lecture Notes in Physics
ISBN 978-3-031-59500-4 ISBN 978-3-031-59501-1 (eBook)
https://doi.org/10.1007/978-3-031-59501-1

© The Editor(s) (if applicable) and The Author(s), under exclusive license to Springer Nature Switzerland AG 2024

This work is subject to copyright. All rights are solely and exclusively licensed by the Publisher, whether the whole or part of the material is concerned, specifically the rights of translation, reprinting, reuse of illustrations, recitation, broadcasting, reproduction on microfilms or in any other physical way, and transmission or information storage and retrieval, electronic adaptation, computer software, or by similar or dissimilar methodology now known or hereafter developed.
The use of general descriptive names, registered names, trademarks, service marks, etc. in this publication does not imply, even in the absence of a specific statement, that such names are exempt from the relevant protective laws and regulations and therefore free for general use.
The publisher, the authors and the editors are safe to assume that the advice and information in this book are believed to be true and accurate at the date of publication. Neither the publisher nor the authors or the editors give a warranty, expressed or implied, with respect to the material contained herein or for any errors or omissions that may have been made. The publisher remains neutral with regard to jurisdictional claims in published maps and institutional affiliations.

This Springer imprint is published by the registered company Springer Nature Switzerland AG
The registered company address is: Gewerbestrasse 11, 6330 Cham, Switzerland

If disposing of this product, please recycle the paper.

Preface

This book is the outcome of pressure on us from colleagues to explain the applications in physics to topology and differential geometry in as simple terms as possible. The book is pitched at a level of the masters and first-year PhD students.

When we started writing this book, we realised the amount of material on topological applications was enormous. Since the background of the students will not be beyond the conventional mathematical physics courses in the Universities, we chose only some topics based on our teaching experience with similar students in our institutions.

The book is divided into three parts. Each part contains three chapters.

The first part introduces the basic notions of topological spaces and their mapping. Here we introduce homeomorphisms, homotopy and homology invariants with examples. Then we follow with differentiable manifolds and differential calculus. After introducing integration on such manifolds we bring out cohomology and connect up with earlier invariants. This is followed by the introduction to Riemannian geometry and Pseudo Riemannian geometry which are essential for general relativity.

The chapters in the second part bring out the simple applications. We review the types of defects in materials and how they are related to topological features of descriptions of crystals. We indicate how the invariants introduced in the earlier part of chapters directly entered are demonstrated. This is followed by simple quantum mechanical examples and profound notions of fundamental statistics of particles emerging from the topology of configuration space. In addition to bosons and fermions which students are aware of, we show the appearance of anyons in two dimensions. Then the fundamental theorem in quantum physics namely the connection between spin and statistics is brought to understand the significance of topology in quantum physics.

After such simple applications, we considered in the last part the topology and geometry of three-dimensional manifolds. We introduced the Chern-Simons gauge theory, which is a topological field theory, and its applications which are of modern interest. After obtaining the knot invariants from this topological gauge theory, we provide the procedure to obtain invariants for some three-dimensional manifolds. Then we elaborate on the physics of three-dimensional black holes in manifolds with a cosmological constant which is negative (known as Anti deSitter spaces). Particularly, we discuss how the Bekenstein-Hawking entropy can be obtained from the same topological theory. This completes our presentation of selective advanced-level applications.

We have provided simple exercises at the end of each chapter. The readers will get a better understanding of the concepts and tools through solving these exercise problems. We will not provide solutions mainly to make sure that every student attempts such problems.

There are many more formal topics like fiber bundles, instantons, skyrmions, integrable models, Hopf algebra and topological aspects of gauge theories which we have not discussed. Also, the applications like Quantum Hall effect (integer and fractional) Hopfions, topological insulators, fuzzy and noncommutative geometry are not presented in this book. Taking up these will need another volume which is for the future.

Chennai, India
Mumbai, India

T. R. Govindarajan
Pichai Ramadevi

Acknowledgements

As we mentioned in the preface there were several colleagues who encouraged us to write this book. We would like to thank them (without specifically naming them, respecting their wish).

Many sections of the contents of this book grew out of our collaborative works with several researchers. We thank all of them. Thanks to them, our knowledge enhanced enormously through interactions with them and we understood nuances during the course of these publications. Specifically we would like to mention Profs. A P Balachandran, Romesh Kaul, Suneeta Varadarajan.

We were guided by the earlier book of Charles Nash, and Siddhartha Sen for Introductory parts.

We were helped during the preparation by Mr Sagnik Banerjee (B Tech student of IIT Bombay) who went through meticulously and offered several comments to make the contents easier for students. Also Mr Sachin Chauhan (PhD student of IIT Bombay) helped us with figures. We record our thanks to them. Lastly TRG would like to thank many faculty members of the Dept of Physics, IIT Bombay who provided lively environment during visits for preparing the book.

Contents

Part I

1 Introduction 3
   1.1 Topology, Maps, Classes, Homotopic Groups 3
      1.1.1 Topological Space 3
      1.1.2 Topological Invariants 8
   1.2 Homotopy Groups 11
      1.2.1 The Fundamental Group... 11
      1.2.2 Quotient Space 17
   1.3 Higher Homotopy Groups 19
   Appendix 21
   Exercises 21
   References 22

2 Differentiable Manifolds and Geometry................................ 23
   2.1 Euclidean Geometry, Metric Spaces, Smooth Manifolds 23
   2.2 Differentiable Manifolds... 24
   2.3 Differential Calculus on Manifolds 25
      2.3.1 Closed and Exact Forms 31
   2.4 Homology 32
   2.5 Simplicial Complex and the Fundamental Group 35
   2.6 Cohomology... 40
      2.6.1 Integration on a Manifold with Boundary 42
   Exercises 46
   References 48

3 Riemannian and Pseudo Riemannian Geometry 49
   3.1 Riemannian Geometry 50
      3.1.1 Inner Product and the Metric 50
      3.1.2 Parallel Transport and Connection 51
   3.2 Riemann Tensor 53
      3.2.1 Geodesics 54
   3.3 Laplace Beltrami Operator 55
   3.4 Self Adjoint Operators 56

		3.4.1 Self-Adjoint Laplace Beltrami Operator on a Manifold with Boundary	58
	3.5	Pseudo Riemannian Geometry	59
		3.5.1 Einstein Hilbert Action	60
		3.5.2 Rindler Spacetime	60
	Exercises		61
	References		63

Part II

4 Topological Understanding of Defects in Crystalline Structure 67
   4.1 Point Defects, Line Defects, and More 67
       4.1.1 Defects and Topology 70
       4.1.2 Group Theory and Order Parameter Space 72
   Appendix: Exact Sequences 75
   Exercises 77
   References 78

5 Configuration Space Topology and Topological Conservation Laws ... 79
   5.1 Paths and Path Connectedness, Homotopy 79
       5.1.1 Quantization Ambiguity and Fundamental Group 80
   5.2 Quantum Mechanical Systems on Different Topological Spaces 81
       5.2.1 A Particle Moving on a Circle $\mathbf{S}^1$ 81
       5.2.2 Particle Confined Within the Interval $[-1, 1]$ 83
       5.2.3 Particle on $\mathbb{R}^1_+$ 84
       5.2.4 Particle Moving in the Space $\mathbb{R}^2 - \mathbf{D}^2$ 84
       5.2.5 Rigid Rotor... 85
   5.3 System of N Identical Particles and Origin of Statistics 86
       5.3.1 Anyons in $d = 2$ Dimensions 88
   5.4 Field Theory on Topological Spaces 90
       5.4.1 Kink Soliton... 90
       5.4.2 $O(3)$ Nonlinear σ Model 93
   Exercises 95
   References 97

6 Spin-Statistics Theorem, Low Dimensional Topology and Geometry 99
   6.1 History 99
       6.1.1 Assumptions 100
   6.2 Anyons and Fractional Spin 102
       6.2.1 $O(3)$ Nonlinear Sigma Model................................. 103
       6.2.2 Interpretation of Hopf Invariant 104
   6.3 Low Dimensional Topology and Geometry.......................... 105
       6.3.1 Genus g Riemann Spaces 105
   6.4 Three-Dimensional Geometry and Topology 107
       6.4.1 Three Manifolds 107

6.5　Chern-Simons Gauge Theory on $\mathbf{S}^3$ 108
Exercises 110
References 111

Part III

7　Braid Group, Knots, Three Manifolds 115
　7.1　Braids, Braid Group 115
　　　7.1.1　Knots 116
　　　7.1.2　Some Examples 117
　　　7.1.3　Knot Invariants 117
　7.2　Chern Simons Theory and Knot/Link Invariants 120
　　　7.2.1　$U(1)$ Chern-Simons Theory and Linking Invariants 121
　　　7.2.2　$SU(2)$ Chern Simons Theory on $\mathbf{S}^3$ 123
　　　7.2.3　Canonical Quantisation 123
　　　7.2.4　Brief Introduction to the $SU(2)_k$ Kac-Moody Algebra 125
　　　7.2.5　Hilbert Space for Flat Connections 126
　7.3　Jones Polynomial from CS Theory 128
　　　7.3.1　Wess-Zumino Witten Models, Quantum Groups,
　　　　　　Vertex Models 131
　　　7.3.2　Vertex Models and Knot Invariants 134
　　　7.3.3　Knot Invariant from Vertex Model 135
　　　7.3.4　Composite Braids and Murakami Invariants 135
　Exercises 137
　References 137

8　Three-Manifold Invariants 139
　8.1　Surgery Procedure 140
　8.2　Kirby Moves 143
　8.3　Three-Manifold Invariants 144
　8.4　Conclusion 147
　Exercises 147
　References 148

9　3D Gravity and BTZ Blackhole 149
　9.1　3D Gravity 150
　　　9.1.1　Killing Symmetries of AdS_3 151
　9.2　Chern Simons Formulation of 3D Gravity 152
　　　9.2.1　BTZ Blackhole 154
　　　9.2.2　BTZ Blackhole and Statistical Mechanics 155
　　　9.2.3　dS_3 Gravity and Chern Simons Formulation 162
　　　9.2.4　Euclidean de Sitter Space in 3D 163
　　　9.2.5　Logarithmic Correction to the Entropy 165
　　　9.2.6　Logarithmic Corrections and AdS/CFT 166
　Exercises 167
　References 167

List of Figures

Fig. 1.1	Associativity of homotopic classes	14
Fig. 1.2	Map from $\mathbf{S}^2 \to Y$	20
Fig. 1.3	Product of maps	20
Fig. 1.4	$\Pi_2(Y)$ is Abelian	20
Fig. 1.5	Winding number N	21
Fig. 2.1	1,2,3-Simplices	33
Fig. 2.2	Deformation to a simplex	34
Fig. 2.3	Polyhedron	34
Fig. 2.4	Triangulation of circle and disc	35
Fig. 2.5	Triangulation of cylinder	35
Fig. 2.6	Triangulation of torus	35
Fig. 2.7	Cylinder triangulation	37
Fig. 2.8	Torus complex with 7 vertices	37
Fig. 2.9	Stereographic projection	45
Fig. 3.1	Rindler spacetime	61
Fig. 4.1	Point defects in a crystal	68
Fig. 4.2	Line defects in a crystal	68
Fig. 4.3	1,2,3-Winding number	71
Fig. 4.4	n+m winding number	72
Fig. 4.5	1-1 winding number	72
Fig. 5.1	Braiding relation 1	89
Fig. 5.2	Braiding relation 2	89
Fig. 5.3	Soliton number one solution	92
Fig. 5.4	Energy density of the soliton	92
Fig. 6.1	Spin statistics	101
Fig. 6.2	Genus g surface	105
Fig. 7.1	Unknot and trefoil	117
Fig. 7.2	Hopf link and figure 8 knot	117
Fig. 7.3	Right and left handed trefoil	118
Fig. 7.4	Borromean rings	121
Fig. 7.5	Manifold M_L and M_R with four-punctured $\mathbf{S}^2$ boundary	128
Fig. 7.6	Markov move 1	134
Fig. 7.7	Markov move 2	134
Fig. 7.8	Composite Braid $B^{(2)}(b_1)$	136

Fig. 8.1	Mathematical curve C in $M = \mathbf{S}^2 \times \mathbf{S}^1$	140
Fig. 8.2	Tubular neighbourhood of C in $M = \mathbf{S}^2 \times \mathbf{S}^1$ '	140
Fig. 8.3	Removed tubular Neighbourhood $\mathbf{N}$ with $\mathbf{T}^2$ boundary	141
Fig. 8.4	Knot complement $M/C = (\mathbf{S}^2 \times \mathbf{S}^1)/C$ with $\mathbf{T}^2$ boundary	141
Fig. 8.5	Framed link(s) and the corresponding three-manifold	142
Fig. 8.6	Kirby moves on any framed link $U(Y)$	143
Fig. 9.1	Euclidean blackhole as solid torus	156
Fig. 9.2	Penrose diagram for dS_3	163

Part I

Introduction 1

1.1 Topology, Maps, Classes, Homotopic Groups

After the developments of Lagrangian and Hamiltonian formulation of the laws of mechanics, we get a clearer picture of the role of symmetries in any physical system. In concrete terms, these symmetries provide *group structure* and conservation laws leading to a better understanding of such systems. However, the role of *topology and geometry* in dynamical systems came to prominence in the last century after the great contributions of Poincare in the book 'Celestial Mechanics' [1] and Birkhoff [2].

To understand the importance of topology and their applications to physical systems, we need to learn from the first principle the definition of the terminologies and notations. For instance, the mathematical terminology like maps, classes, homeomorphism, homotopy, homology, etc will be discussed along with simple examples in this chapter.

1.1.1 Topological Space

While our interest will be for manifolds with differential structures, we will start from topological space which can also be defined for a discrete set of elements [3]. Readers interested in the physics approach can refer to the book by Nash and Sen [4].

Let us begin with a set X and a collection of subsets $Y = \{X_i\}$. This collection could be finite or infinite in number. Now we define a topological space (X, Y) if $\{X_i\}$'s satisfy a set of conditions.

1. The null set $\emptyset$ and complete set X are contained in Y.
   i.e., $\emptyset \in Y$, and $X \in Y$.

2. Union of sub-collection of subsets $\{Z_j\} \subset \{X_i\}$ is also contained in Y.
   i.e., $\bigcup_j Z_j \in Y$. Here, the sub-collection $\{Z_j\}$'s could be a finite or infinite collection of subsets.
3. Any intersection of finite sub collection $\{Z_j\}$ of $\{X_i\}$ is also contained in Y.
   i.e., $\bigcap_{j=1}^{n} Z_i \in Y$.

The above three properties define the set X as a topological space whose topology is provided through the choice Y. The subsets $X_i \in Y$ are called open sets.

Examples

1. If $Y = \{\emptyset, X\}$, then the conditions are trivially satisfied. Such a Y denotes the trivial topology of X.
2. If Y contains all possible subsets of X then again the conditions are automatically satisfied. This is known as the discrete topology of X
3. Consider $\mathbb{R}$ denoting the set of real numbers. Consider an open set (a, b) containing $x \in \mathbb{R}$. That is, $a < x < b$. If $Y = \{(a, b)\}$, with $a, b \in \mathbb{R}$ along with their unions, then the conditions are satisfied. This is the conventional topology of $\mathbb{R}$ which can be identified as an infinite elastic string.

To motivate such conditions above and relate to the conventional notions of elastic sheet topology, we need formal definition of continuous functions.

Definition 1.1 (Function) A function f is a mapping from the topological space (X, Y) to another topological space (U, V):

$$f : X \to U .$$

Definition 1.2 (Continuous Function) A function f is continuous if the inverse image of an open set in U is an open set in X.

Let us explain the above definitions for the function $f : \{X = \mathbb{R}\} \to \{U = \mathbb{R}\}$.
As an example, consider the following function:

$$f(x) = \theta(x) .$$

That is $f(x) = 0$, $x \leq 0$, $f(x) = 1$, $x > 0$. Clearly this function will have a discontinuity at $x = 0$. But we will see how this emerges from our definition of continuous functions.

In this example, $f(x \in X) \in \{0, 1\}$. Hence, all possible open subsets:

$$\emptyset, \{0\}, \{1\}, \{0, 1\}$$

defines the discrete topology for the topological space (U, V). Clearly the inverse image

$$f^{-1}(\{1\}) = (0, \infty) \in X$$

is open. However, the inverse image for the other open subset

$$f^{-1}(\{0\}) = (-\infty, 0]$$

is not an open set. It is also interesting to note that actually the function $f[X_i = (a, b)]$ maps open sets $X_i \in Y$ to open sets belonging to discrete topology V. It is the inverse map $f^{-1}(U_i \in V)$, where U_i is an open set, is not necessarily an open set. Hence violates the continuity condition thereby making the function discontinuous.

It is important to realise that the definition of continuity requires the finite intersection of open sets to be open sets but not the infinite intersection of open sets. In the above example, the two points $p \in U$ ($p = \{0\}$ or $\{1\}$) are open sets in discrete topology V which can be obtained as an infinite intersection of open sets

$$\bigcap_{i=1}^{\infty}(p - a_i, p + a_i)$$

where $a_i \in \mathbb{R}$. This makes the condition too rigid for continuity.

Now, we provide further definitions which are required to develop notions of topological space.

Definition 1.3 (Closed Set) A subset $S \in X$ in a topological space X is a closed set if the complement viz., $S^c = X - S$ is an open set.

Corollary 1.1 *Note a set is open if the complement is closed. Further, the null set $\emptyset$ and the total space X are both open and closed.*

Definition 1.4 (Neighbourhood) A neighbourhood of a point 'x' in a topological space X is **N**, if x is contained in an open set X_i which is contained in **N**:

$$x \in X_i \subseteq \mathbf{N}$$

Note 1.1 **N** *itself may not be an open set but should contain an open set which includes x.*

Consider, for example, $\mathbb{R}$ with the usual topology. The neighbourhood of a point $x = 2$ can be $=[1, 3]$. Here, the neighbourhood is not an open set. If the neighbourhood is an open set, it is called an open neighbourhood.

Corollary 1.2 (Intervals) *Consider an interval $[x, y]$. This is easily seen to be complement of $(-\infty, x) \cup (y, \infty)$. Hence, any closed interval is closed.*

Definition 1.5 (Closure) Closure of a subset $S \subseteq X$ for the topological space (X, Y) is denoted as $\bar{S}$ and defined as

$$\bar{S} = \{x \in X : \forall X_i \in Y \text{such that } x \in X_i \text{ and } X_i \cap S \neq \emptyset\}$$

We can easily see the closure of the set S is the smallest closed set $\bar{S}$ containing S. For the simplest example $X = \mathbb{R}$, if the open set $S = (a, b)$ then the closure is $\bar{S} = [a, b]$. If we start with a closed set then closure is itself. Closure is useful in defining the boundary and interior of a set which we discuss now.

Definition 1.6 (Interior) Interior S^0 of a set S is the union of all the open subsets $S_i \subseteq S$:

$$S^0 = \bigcup_i S_i .$$

The boundary $\mathbf{B}(S)$ of a set S is the complement of the interior of S in the closure of S:

$$\mathbf{B}(S) = \bar{S} - S^0 .$$

Note 1.2 *Interior of a set is the largest open set contained in S.*

Examples

1. Take $\mathbb{R}^3$. Consider a ball defined by $x_1^2 + x_2^2 + x_3^2 \leq 1$. Here, the interior would be the interior of the ball excluding the surface $\mathbf{S}^2$ of unit radius. The boundary would be the unit sphere $x_1^2 + x_2^2 + x_3^2 = 1$.
2. Consider the subsets of rationals S in $\mathbb{R}$. That is., those numbers which can be expressed as $\mathbb{Q} = \frac{p}{q}$ where p, q are integers. This subset is disconnected. Only connected subsets of $\mathbb{R}$ are:

$$[a, b], (a, b), [a, b), (a, b], (-\infty, a], (-\infty, a), (a, \infty), [a, \infty), (-\infty, \infty).$$

In every open interval, however small it is, we have infinitely many rational numbers. The boundary $\mathbf{B}(\mathbb{Q})$ of $\mathbb{Q}$ contains $\mathbb{Q}$. The interior of $\mathbb{Q}$ is a null set $\emptyset$ because all the subsets of $\mathbb{Q}$ are closed. Hence $\mathbf{B}(\mathbb{Q}) = \mathbb{R}$, $\bar{\mathbb{Q}} = \mathbb{R}$. That is closure of $\mathbb{Q}$ is $\mathbb{R}$ itself. This makes S as dense subset of $\mathbb{R}$ which is defined as follows:

Definition 1.7 (Dense Set) A subset S of X is said to be dense if the closure $\bar{S}$ of S is X itself.

We will now consider some examples to understand whether the topological space is 'compact' or 'non-compact.' The simplest real line $\mathbb{R}$ has infinite length and is obviously non compact topological space. Also the set (a, ∞) has infinite length and hence not compact. Clearly, all topological spaces of infinite volume will not be compact. On the other hand, we need further condition for a topological space to be compact even if it has finite volume. We will provide definition of 'cover' which will lead to the concept of 'compact topological space'.

Definition 1.8 (Cover) We want to cover a set S by a collection of subsets $\{F_i\}$. We define $\{F_i\}$ is a cover of S if $\bigcup F_i \supseteq S$. If the $\{F_i\}$ are all open sets then the cover is known as open cover.

We now use the above open cover to define 'compact spaces.'

Definition 1.9 (Compact Space) A space S is compact if there exists a finite set of open sets $\{F_i\}, i = 1, \ldots N$ for every open covering such that $\bigcup_{i=1}^{N} F_i \supseteq S$.

Compactness generalises the concept of closed subset which is bounded. Examples are: $(-\infty, a]$ is not compact since it is not bounded—i.e., there exists points separated by infinite distance. The interval (a, b) is not compact since it is not closed. That is limit of sequence of points will reach 'a' or 'b' which are not included in the set. Lastly the interval $[a, b]$ is compact since it is bounded and closed. That is the distance between any two points is finite and the sequence of points end with 'a' or 'b' which are included in the set. In two dimensions $\mathbb{R}^2$, closed discs defined as

$$\{(x, y) : x^2 + y^2 \leq R^2\}$$

are compact but open discs

$$\{(x, y) : x^2 + y^2 < R^2\}$$

are not. Similarly, spheres are compact whereas a point removed from sphere ($\mathbf{S}^2 - \{point\}$) is not compact.

Note 1.3 *A subset of a compact space can be non-compact.*

For example $(1, 2]$ is not compact, but $[1, 2]$ is. Set of rational numbers in $[0, 1]$ is non compact!. This is because the intervals $\left[0, \frac{1}{\pi} - \frac{1}{n}\right] \cup \left[\frac{1}{\pi} + \frac{1}{n}\right], \forall n = 4, 5, ..$ covers the interval, but there are no finite sub covers.

Even though our discussions uses $\mathbb{R}^n$ as examples for compact spaces, these can be generalised. There could be some novel features emerging when other general spaces are considered but physical intuitions from $\mathbb{R}^n$ examples are sufficient to understand them.

Before moving on to discussions about homeomorphism and homomorphism, which are required to understand our elastic sheet model of topological spaces, we will complete this subsection with the notion of connected and disconnected spaces.

Definition 1.10 (Connected Space) A topological space S is disconnected if it can be written as $S = S_1 \cup S_2$ where S_1, S_2 are both open sets and $S_1 \cap S_2 = \{\emptyset\}$. If such a division is not possible then the space is connected.

For example in $\mathbb{R}$, a set $(a, b) \oplus (c, d)$ is disconnected if both the sets are not overlapping which means $(a, b) \cap (c, d) = \{\emptyset\}$. This can be easily generalised to $\mathbb{R}^n$ too. If we can draw disconnected pieces then the space will be disconnected.

1.1.2 Topological Invariants

We imagine the topological space as an elastic medium with allowed distortions without tearing the space. This activity allows us to go from one space to another continuously. This is achieved by introducing the framework of homeomorphism.

Definition 1.11 (Homeomorphism) If we have two topological spaces X, Y and a map f from X to Y

$$f : X \longrightarrow Y \qquad (1.1)$$

is a homeomorphism if it is continuous and has an inverse f^{-1} which is also continuous.

Note that, if f is a homeomorphism, then f^{-1} is also. Homeomorphism is transitive. That is., if X is homeomorphic to Y and Y is homeomorphic to Z, then by composing the two maps we will have X is homeomorphic Z. This brings us to the idea of equivalence classes. We can put different topological spaces in different classes. Two spaces in the same class are homeomorphic to each other whereas the two in different classes are not homeomorphic to one another.

We can now ask the question of how we characterize each of the classes. We need to define topological invariants which will provide distinguishing features on different topological spaces. That will help us to classify the topological spaces completely. But this is possible in low-dimensional spaces like one, two and three

1.1 Topology, Maps, Classes, Homotopic Groups

dimensions. In principle dimension of the topological space is an invariant. We cannot deform spaces of different dimensions to one another. We can also identify compactness as another invariant. A compact space cannot be deformed to a non-compact one. Similarly, connectedness is also another feature of the topological class. Two spaces with differing connectedness cannot be mapped to one another continuously.

Note 1.4 *Two spaces with differing topological invariants cannot be homeomorphic to each other.*

Dimension, Compactness, Connectedness as Invariants

We will establish from the first principle that a compact space cannot be homeomorphic to a non-compact space. Let us start spaces X, Y where X is compact and Y is non-compact. Let us assume f is a map.

$$f : X \to Y \tag{1.2}$$

Let $\{F_i\}$ be the open cover of Y. Since f is a continuous map, we have $f^{-1}(F_i)$ is an open set of X. As X is compact we have a finite subcover $f^{-1}(\{F_i\})$. This means a finite subcover of Y also. But Y is non-compact and hence, such a map cannot exist.

If a space X is connected and Y is not connected, then we will show that there cannot be a homeomorphism between them. As Y is disconnected, we can write

$$Y = Y_1 \cup Y_2, \quad Y_1 \cap Y_2 = \{\emptyset\}$$

and Y_1, Y_2 are open. Since f is continuous, $f^{-1}(Y_1)$, $f^{-1}(Y_2)$ are open in X. They also obey $f^{-1}(Y_1) \cup f^{-1}(Y_2) = X$. But that would imply X is not connected. Hence, the map must have an obstruction.

We mentioned that the dimension is an invariant. We will provide further arguments to substantiate the statement:

- Consider $\mathbb{R}$ and $\mathbb{R}^2$ and a map

$$f : \mathbb{R} \to \mathbb{R}^2 .$$

If $\mathbb{R}$ and $\mathbb{R}^2$ are not homeomorphic, then there should be an obstruction to this map. Consider $\mathbb{R}^2$ and remove the origin denoted by a point $(0, 0) \in \mathbb{R}^2$. Such a space is still connected. However, the point '0' removed from $\mathbb{R}$ makes the space $\mathbb{R} - \{0\}$ disconnected with two spaces. That means, $\mathbb{R} - \{0\}$ is not homeomorphic to $\mathbb{R}^2 - \{0, 0\}$.

- Consider the map f restricted as follows:

$$f : \mathbb{R} - \{0\} \to \mathbb{R}^2 - \{0, 0\} . \tag{1.3}$$

We have already discussed that it is impossible for the restricted map to be homeomorphic.

This argument can be extended to maps from $\mathbb{R}^2$ to $\mathbb{R}^3$ and so on. Hence we can conclude two spaces $\mathbb{R}^n$ and $\mathbb{R}^m$ can never be homeomorphic unless $n = m$.

Going further we can associate a mathematical structure like a group to topological spaces. These structures are homotopy groups, homology groups, isotopy groups, cohomology groups etc. We will now define homotopy groups as it is a powerful ingredient to characterise the topological classes.

Definition 1.12 (Homotopy) Two maps f_1, f_2 from the space X to Y

$$f_1 : X \longrightarrow Y, \qquad f_2 : X \longrightarrow Y \tag{1.4}$$

are said to homotopic to each other ($f_1 \approx f_2$) if f_1 can be deformed to f_2 continuously.

This can be stated precisely: There exists a one parameter family of maps $F(t)$ from $F : X \longrightarrow Y$ such that

$$F(0) = f_1, \qquad F(1) = f_2 \tag{1.5}$$

Homotopy is also transitive, which means if $f_1 \approx f_2$ and $f_2 \approx f_3$, then $f_1 \approx f_3$. This provides us with equivalence classes of homotopic maps.

We have already shown that the homeomorphism provides us equivalence classes whose elements are topological spaces and we see now that the homotopy gives equivalence classes whose members are maps. Homeomorphism is also a continuous map which implies classes of homotopic maps are unaltered under homeomorphism of X or Y. These homotopy equivalence classes are topological invariants of the pair of spaces X and Y. We denote these homotopy classes between X and Y by the symbol $[X, Y]$. This implies, if the homotopy classes are different between two pairs of spaces then such spaces cannot be homeomorphic to each other.

In fact, we will see that the homotopy invariants can be used to classify homeomorphism classes. For that purpose we fix one of the space to be always the same, say for example $X = \mathbf{S}^n$ the round sphere in n-dimension. We consider different spaces Y_i and compare the homotopy classes $[\mathbf{S}^n, Y_i]$. If two different spaces Y_1 and Y_2 have different homotopy classes $[\mathbf{S}^n, Y_1]$ and $[\mathbf{S}^n, Y_2]$ for some 'n', then we can claim that the two spaces Y_1 and Y_2 cannot be homeomorphic as there will be some obstruction to the maps.

The obstructions between different classes of maps characterise the topological properties of the space Y itself. We denote the classes of maps $[\mathbf{S}^n, Y]$ by $\Pi_n(Y)$. We will now establish the homotopic classes have a natural group structure and are known as homotopy groups.

1.2 Homotopy Groups

In the previous section we introduced the framework of homotopy classes $\Pi_n(Y) = [S^n, Y]$. We now study these classes and the group structure amongst them for different n. Let us consider the simplest case of $n = 1$ to understand the concept of a fundamental group.

1.2.1 The Fundamental Group

We begin with the following definitions as a prelude to discuss the fundamental group.

Definition 1.13 (Path) A path $p(t)$ is a continuous map from the closed interval $[0, 1]$ in $\mathbb{R}$ to Y such that $p(t = 0) = y_0 \in Y$ and $p(t = 1) = y_1 \in Y$. That is,

$$p(t) : [0, 1] \longrightarrow Y, \qquad p(0) = y_0, \ p(1) = y_1. \tag{1.6}$$

A topological space is path connected if there exists a path between any two arbitrary points. We should remind ourselves of the earlier definition of connected topological space. For most topological spaces which we study in physical systems, there always exists a path between any two points. In that sense a path connected space and connected space are one and the same.

Definition 1.14 (Loop) A loop or a closed path is a continuous map $p(t)$ for which the starting and ending points are the same, i.e. $y_0 = y_1$. Loop is

$$p(t) : [0, 1] \longrightarrow Y, \qquad p(0) = p(1) = y_0. \tag{1.7}$$

Now to develop the group structure, these loops based at a point y_0 can be put in equivalent classes. We will now present the definitions to check whether any two loops $p_1(t)$, $p_2(t)$ are in the same class or not,

Definition 1.15 (Equivalence Class) Two loops are in the same class if there exists a one-parameter family of loops $p(t, s)$ such that for all $p(t, s)$ with $p(0, s) = p(1, s) = y_0$ we have $p(t, 0) = p_1(t)$, and $p(t, 1) = p_2(t)$

Physically, this means we can smoothly interpolate between the two loops without facing any obstruction and hence belong to the same equivalent class. This implies $p_1(t)$ is in the same equivalence class of $p_2(t)$. We will denote the equivalent classes by $[p_1(t)]$. Since our goal is to obtain a group structure of such equivalence classes of loops, we define a product rule between any two loops.

Definition 1.16 (Product of Loops) The product $p_3(t)$ of two loops $p_1(t)$ and $p_2(t)$ at the same base point ($p_1(0) = p_1(1) = p_2(0) = p_2(1)$) is defined as

$$\begin{aligned} p_3(t) &= p_1(t) \otimes p_2(t) \\ &= p_1(2t), \quad 0 \leq t \leq \frac{1}{2} \\ &= p_2(2t-1), \quad \frac{1}{2} \leq t \leq 1 \end{aligned}$$

The above definition implies the product is constructed first traversing along the loop $p_1(t)$ and then following it up with $p_2(t)$. There is a conventional difference between physics and mathematics literature on this by the interchange of the order. Secondly the rule can be extended to any member of the classes $[p_1(t)], [p_2(t)]$. From the product of loops, we can define the inverse of a loop and the identity loop.

Definition 1.17 (Inverse of a Loop) The inverse of a loop $p(t)$ based at y_0 will traverse in the opposite direction, i.e.,

$$p^{-1}(t) = p(1-t). \quad 0 \leq t \leq 1 \tag{1.8}$$

Again, we can extend the above to any member of the class. For this set of homotopy classes to satisfy group properties, one needs the 'identity' loop.

Definition 1.18 (Identity) Identity loop $I(t)$ is given by

$$I(t) = y_0, \quad 0 \leq t \leq 1 \tag{1.9}$$

All loops which are continuously contractible to a point as defined above are also in the class of Identity loop.

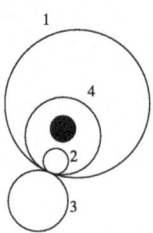

In this diagram, the blackened spot refers to a hole in $\mathbb{R}^2$. Various loops are drawn based at a point $y_0 \in \mathbb{R}^2$. Clearly, the loops numbered 1 and 4 are homotopic to each other. Hence, they belong to an equivalent class. Similarly, the loops 2 and 3 are homotopic to each other and can be shrunk to a point. Hence, loops 2 and 3 are trivial denoting the identity loops, But loops 1 and 4 are not homotopic to loops 2 and 3 due to the obstruction of the hole.

1.2 Homotopy Groups

Given the above definitions, we can establish that the equivalent classes of loops based at a point $y_0 \in Y$ form a group. This group is called the Fundamental group and is denoted by $\Pi_1(Y) = [\mathbf{S}^1, Y]$. For completeness, we have given a brief introduction to discrete groups in Appendix at the end of this Chapter.

Theorem 1.1 (Fundamental Group) *With the definition of product rule between equivalent classes and the inverse of loops, the set of homotopic classes of loops form a group $\Pi_1(Y, y_0)$. The identity element refers to all the loops that are contractible to a point.*

We have already shown the product of two loops leads to another loop, which is also in the set. The product of a loop with its inverse will give an identity element. The associativity of the product of three loops needs to be shown, to fulfill group structure, which we provide now.

Associativity of three loops p_1, p_2, p_3 is:

$$p_1 \otimes (p_2 \otimes p_3) = (p_1 \otimes p_2) \otimes p_3 \tag{1.10}$$

We can write the LHS as:

$$\begin{aligned} p_1 \otimes (p_2 \otimes p_3) &= p_1(2t), & 0 \leq t \leq 1/2 \\ &= p_2(4t - 2), & 1/2 \leq t \leq 3/4 \\ &= p_3(4t - 3), & 3/4 \leq t \leq 1 \end{aligned}$$

In the same way we can write the RHS:

$$\begin{aligned} (p_1 \otimes p_2) \otimes p_3 &= p_1(4t), & 0 \leq t \leq 1/4 \\ &= p_2(4t - 1), & 1/4 \leq t \leq 1/2 \\ &= p_3(2t - 1), & 1/2 \leq t \leq 1 \end{aligned}$$

LHS and RHS can be shown to be in the same class by introducing $p(t, s)$ such that $p(t, s = 0) = $ RHS and $p(t, s = 1) = $ LHS:

$$\begin{aligned} p(t, s) &= p_1\left(\frac{4t}{s+1}\right), & 0 \leq t \leq \frac{s+1}{4} \\ &= p_2(4t - s - 1), & \frac{s+1}{4} \leq t \leq \frac{s+2}{4} \\ &= p_3\left(\frac{4t - 2 - s}{2 - s}\right), & \frac{s+2}{4} \leq t \leq 1 \end{aligned}$$

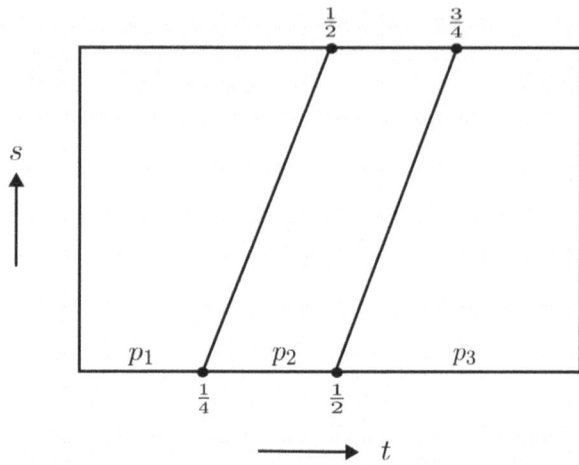

Fig. 1.1 Associativity of homotopic classes

Such an associativity property is graphically depicted in Fig. 1.1. Note that, as we change s from 0 to 1 we map RHS to LHS. We need to check next whether the fundamental group depends on the base point y_0. We will now show that $\Pi_1(Y, y_0)$ and $\Pi_1(Y, y_1)$ are isomorphic.

Construct a path α from y_0 to y_1. The reverse path will be α^{-1}. We can easily convert a loop $p(t)$ based at point y_0 to a loop based at y_1 by the following product of paths: $\alpha^{-1} \otimes p(t) \otimes \alpha$. In the similar way, we can convert loops at y_1 to those at y_0 as well. Hence we get

$$\Pi_1(Y, y_0) \cong \Pi_1(Y, y_1). \tag{1.11}$$

Caution Even though the two groups are isomorphic there is no canonical isomorphism between them. That is, we can change the loop at y_0 to that at y_1 by introducing another loop passing through y_0 and y_1 as shown:

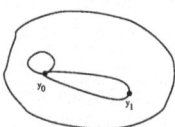

As a result, the maps belonging to the homotopic classes based at y_0 will be different from the homotopy classes based at y_1.

Is the fundamental group a topological invariant? To answer this question, we first introduce the definition of same homotopy type spaces X, Y.

1.2 Homotopy Groups

Definition 1.19 (Homotopy Type) Two spaces X, Y are same homotopy type $X \approx Y$ if there are continuous maps f, g such that $f : X \longrightarrow Y$ and $g : Y \longrightarrow X$ such that

$$f \bullet g : X \to X$$

must be homotopic to the identity map but need not be equal to the identity map.

This makes two homeomorphic spaces are the same homotopy type but the converse need not be. For example, $\Pi_1(S^1) = \Pi_1(\text{annulus})$ but S^1 is not homeomorphic to annulus.

Corollary 1.3 (Topological Invariant) *If X and Y are homeomorphic spaces then*

$$\Pi_1(X, x_0) = \Pi_1(Y, y_0) \tag{1.12}$$

This corollary proves fundamental group is a topological invariant.

Having discussed the fundamental group of a topological space we will now explain concept of deformation retract which is useful in obtaining the fundamental group of a topological space.

Definition 1.20 (Retract) Retract is a continuous mapping from a topological space X into a subspace X_r such that it preserves the position of all points in that subspace. The subspace X_r is then called a retract of the original space.

'r' is a retraction if the map:

$$r : X \longrightarrow X_r, \qquad r(x_r) = x_r, \quad x_r \in X_r \tag{1.13}$$

The retract is an identity map on the subspace. We can call such a subspace as an inclusion in the topological space X and indicated as follows:

$$i : X_r \hookrightarrow X, \tag{1.14}$$

such that $r \circ i = id_{X_r}$.

Definition 1.21 (Deformation Retract) A continuous map

$$F : X \times [0, 1] \longrightarrow X \tag{1.15}$$

is a deformation retraction of a space X onto a subspace $A \subset X$ if

$$F(x, 0) = x, \quad F(x, 1) \in A, \quad \text{and} \quad F(a, 1) = a \tag{1.16}$$

for every $x \in X$ and $a \in A$,

Note 1.5 *There could be maps which are retract but not deformation retract.*

From the above, we can see that the deformation retraction is a homotopy between a retraction and the identity map on X. The subspace A is called a deformation retract of X. In fact, these deformation retractions are special cases of the homotopy equivalences.

> **Theorem 1.2** *If Y is deformation retract of X which is path connected then the $\Pi_1(X, x)$ is isomorphic to $\Pi_1(Y, y)$.*

1. Using this theorem we can establish that $\Pi_1(\mathbb{R}^n)$ is isomorphic to that of a point, which is trivial. This is because a single point is deformation retract of $\mathbb{R}^n$.
2. A closed unit disc with origin removed $\mathbf{D}^n - 0$ has deformation retract as $\mathbf{S}^{n-1}$. The following map establishes this:

$$F : (\mathbf{D}^n - \{0\}) \times [0, 1]) \to (\mathbf{D}^n - \{0\}), \qquad (1.17)$$

where

$$F(x, t) = (1 - t)x + t\frac{x}{|x|}, \, x \in \mathbb{R}^n, \, |x| = \sqrt{x.x}. \qquad (1.18)$$

For example $\Pi_1(\mathbf{D}^3 - \{0\}) = \Pi_1(S^2)$. Clearly any loop drawn along a latitude of the sphere $\mathbf{S}^2$ can be shrunk to trivial loop by moving to the north or south pole. Hence the fundamental group $\Pi_1(\mathbf{S}^2)$ is trivial. For $n = 2$, we see the isomorphism between fundamental group of annulus and circle:

$$\Pi_1(\mathbf{D}^2 - \{0\}) = \Pi_1(\mathbf{S}^1).$$

We will elaborate in the following section that the fundamental group $\Pi_1(\mathbf{S}^1)$ is a set of integers.

Definition 1.22 (Simply Connected Space) A topological space with trivial fundamental group is known as a simply connected space.

We list some examples to illustrate the fundamental groups for many topological spaces:

1. It is easy to see $\Pi_1(\mathbf{R}^n) = e$ (trivial group) because all loops are homotopic to identity loop.
2. $\Pi_1(\mathbf{S}^1) = \mathcal{Z}$ the set of integers. We are looking for homotopic classes of maps $\mathbf{S}^1 \to \mathbf{S}^1$. If we label the two circles by θ, ϕ then the loops can be classified under maps $\theta = n\phi$, $n = 0, \pm 1, \pm 2, \ldots$. This integer n is known as winding

number as a loop parameterized by ϕ is traversed n times on $\mathbf{S}^1$ parametrized by θ. Remember the group $\mathcal{Z}$ of integers can be presented through one generator g and all elements are given by $\underbrace{g * g * \ldots g}_{n}$ with $n = 0, \pm 1, \pm 2, \ldots$ where the group operation $*$ is addition. Two loops with $n \neq m$ are not homotopic to each other and naturally belong to different classes.
3. $\Pi_1(\mathbb{R}^2 - \{0\}) = \mathcal{Z}$. The loops can be seen to be to be going either clockwise or anticlockwise around the origin $\pm n$ times which classify the maps.
4. As mentioned earlier, the fundamental group $\Pi_1(\mathbf{S}^2) = e$ (trivial group) as latitudes can be taken to the poles and contracted trivially.
5. The fundamental group of a cylinder $\mathbb{R}^1 \times \mathbf{S}^1$ is again $\mathcal{Z}$.
6. The above example of the cylinder can be generalised for product spaces as $\Pi_1(X \times Y) = \Pi_1(X) \times \Pi_1(Y)$.
7. The fundamental group of $\mathbb{R}^2 - \{(0, 0)\} - \{(0, 1)\}$ can be presented through two generators g_1, g_2 as all possible 'words' $g_1^{n_1} g_2^{n_2} g_1^{m_1} g_2^{m_2} \ldots$. There is no relation between g_1 and g_2. This is known as the free group in two variables. Such a group is also the fundamental group of two touching circles.

Many topological spaces can be obtained as quotient spaces under a group action on simply connected spaces which we elaborate now. Such an action facilitates obtaining the fundamental group of such quotient spaces.

1.2.2 Quotient Space

Given a topological space which is simply connected, there is a simple procedure to obtain a multiply connected space with a nontrivial fundamental group.

Consider a space X which is simply connected. Let a group G act on this space freely. By free action, we mean there is no point $x \in X$ which is left undisturbed by its action. That is, the action of group elements $g \in G$ on any point x gives a different point x' in X.

$$g \bullet x = gx$$

where $x \neq gx$ for all group elements g except the identity element. Now we identify these different points as one and the same point. This procedure is known as quotienting and the resultant space is denoted by X/G. Then the fundamental group of the quotient space is G.

Recall the properties of a simply connected topological space X. All loops are contractible and homotopic to the identity map from $S^1 \to X$. Now the elements $g_i \in G$, $i = 1, 2.., n$ acting on the space X maps a point x to $g_i x$. Identify these points to obtain the quotient space X/G. Now the paths $p_i(t) \in X$ such that

$$p_i(0) = x, \quad p_i(1) = g_i x$$

become loops in X/G. These are not homotopic to identity map. The homotopic classes of maps are in one-to-one correspondence with the elements g_i of the group G. Hence the homotopic class of maps $[S^1, X/G]$ are isomorphic to G. We will present some examples to illustrate the fundamental group of quotient spaces.

1. Consider the space $\mathbb{R}$ of real numbers. On this there is an action of the group of integers $\mathcal{Z}$ through $x \approx x + \mathbb{N}$. In fact, identifying these points and quotienting provides us the space $\mathbf{S}^1$. That is

$$\mathbf{S}^1 = \mathbb{R}/\mathcal{Z}.$$

   Hence the fundamental group of S^1 is $\Pi_1(\mathbb{R}/\mathcal{Z} \equiv \mathbf{S}^1) = \mathcal{Z}$. We call $\mathbb{R}$ as universal covering space of S^1.

2. Consider the space of 2×2 unitary matrices U with $\mathrm{Det} U = 1$. This collection of unitary matrices belongs to a special unitary group called $SU(2)$. Further, we can show that parametrisation of the matrix elements make it isomorphic to $\mathbf{S}^3$. That is, a general element is labelled by two complex number z_1, z_2:

$$U = \begin{pmatrix} z_1 & z_2 \\ -z_2^* & z_1 \end{pmatrix}$$

   obeying $|z_1|^2 + |z_2|^2 = 1$.

   On this group there is natural action by $\mathbf{Z}_2 = e, a;\ a^2 = e$:

$$\mathbf{Z}_2 : U \to -U .\tag{1.19}$$

   Hence we get quotient space as $SU(2)/\mathbf{Z}_2 \equiv \mathbf{S}^3/\mathbf{Z}_2$ whose fundamental group is $\mathbf{Z}_2$ as we know $\mathbf{S}^n$ is simply connected for all $n \geq 2$. Such a quotient space is nothing but the well studied special orthogonal group $SO(3)$. These are used extensively in the angular momentum theory in Quantum Mechanics.

3. The space $\mathbf{RP}^n$ is a real projective space defined as the space of rays in $n + 1$ dimensions. That is, the space of rays in $\mathbb{R}^{n+1} - (0, 0, 0, \ldots 0)$ is obtained by identifying

$$(x_1, x_2, \ldots x_{n+1}) \approx \lambda(x_1, x_2, \ldots, x_{n+1}) ,\tag{1.20}$$

   for all real λ to get $\mathbf{RP}^n$.

   For $n = 1$, the projective $\mathbf{RP}^1$ is to associate a point at unit distance from the origin which takes care of the above identification (1.20) on (x, y) plane with origin removed. Hence, the space $\mathbf{RP}^1$ is a circle $\mathbf{S}^1$ with antipodal (diametrically opposite) points identified. Similarly, for $\mathbf{RP}^2$ is the sphere $\mathbf{S}^2$ with antipodal points identified. That is., $\mathbf{RP}^2 = \mathbf{S}^2/\mathbf{Z}_2$. Hence, the fundamental group of $\mathbf{RP}^2$ from such a quotienting group $\mathbf{Z}_2$ implies

$$\Pi_1(\mathbf{RP}^2) = Z_2 \,. \tag{1.21}$$

From these two simple examples, it is straightforward to visualise $\mathbf{RP}^3$ to be $\mathbf{S}^3$ with antipodal points identified:

$$\mathbf{RP}^3 = \mathbf{S}^3/\mathbf{Z}_2 \,.$$

Such a space is $SO(3)$ as discussed in the previous example.

After this brief introduction to the fundamental group in this section, we will now focus on the higher homotopy groups.

1.3 Higher Homotopy Groups

Now we shall define and construct properties of the higher homotopy groups $\Pi_n(X)$. They are defined by the homotopy class of maps $[\mathbf{S}^n, X]$. Define the surface in $\mathbb{R}^{n+1}$ by $\sum_1^{n+1} x_i^2 = 1$. Such a surface defines n-sphere $\mathbf{S}^n$. For example, $x_1^2 + x_2^2 = 1$ is a circle and $x_1^2 + x_2^2 + x_3^2 = 1$ is a sphere $\mathbf{S}^2$.

Interestingly when $n = 0$, we get $x_1^2 = 1$ which represents two points $x = \pm 1$ which are boundaries of a closed interval $[-1, 1]$. In fact $n = 0$ provides definition for path connectedness of the topological space Y. Hence $[\mathbf{S}^0, Y] = \Pi_0(Y)$ gives the number of path-connected components.

The homotopy class of maps $[\mathbf{S}^2, Y]$ is also known as the second homotopy group denoted by $\Pi_2(Y)$. We can get maps from $\mathbf{S}^2$ by mapping $\alpha(t, s)$ of unit square disc $0 \leq t, s \leq 1$ with the edges of the square identified (see Fig. 1.2). That is:

$$\alpha(0, s) = \alpha(1, s) = \alpha(t, 0) = \alpha(t, 1) = y_0.$$

Two maps $\alpha_1(t, s)$ and $\alpha_2(t, s)$ will be homotopic to each other if one can continuously go from one to another. Figure 1.3 explains the procedure for the product of maps: We can define a product among the classes:

$$\alpha_2 \bullet \alpha_1(t, s) = \alpha_1(2t, s), \ 0 \leq t \leq \frac{1}{2}$$
$$= \alpha_2(2t - 1, s), \ \frac{1}{2} \leq t \leq 1,$$
$$\alpha^{-1}(t, s) = \alpha(1 - t, s)$$

Fig. 1.2 Map from $S^2 \to Y$

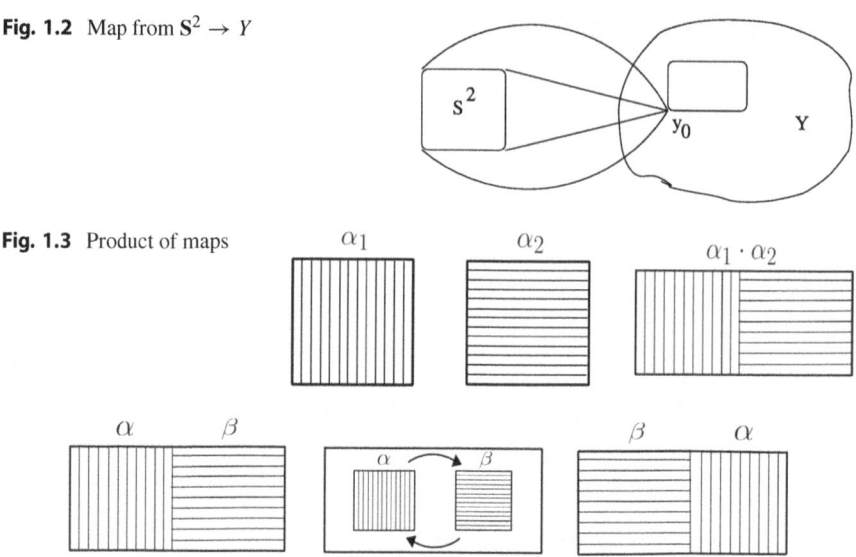

Fig. 1.3 Product of maps

Fig. 1.4 $\Pi_2(Y)$ is Abelian

Hence the classes of such maps can be given a group structure. As the boundary of the square is mapped to y_0, the horizontal and vertical lines $\alpha_1(t, s), \alpha_2(t, s)$ are y_0.

We can move the horizontal and vertical strips continuously such that parametrisation is interchanged. This exhibits diagrammatically that the product of the two maps commute. Such a group property is depicted in Fig. 1.4 exhibiting abelian nature.

This implies all the elements commute among themselves making the $\Pi_2(Y)$ an abelian group. Contrast this with $\Pi_1(Y)$ which can be non-abelian. This difference between $\Pi_1(Y)$ and $\Pi_2(Y)$ arises because the boundary of the interval used for Π_1 is not a path connected space (two points), but the boundary of the square disc is a circle and is a path connected topological space.

We can extend the above analysis to $\Pi_n(Y), n \geq 2$ by considering higher dimensional hypercubes. All of them by similar analysis are abelian. Only $\Pi_1(Y)$ can be non-abelian.

It is easy to understand that $\Pi_2(S^2) = [S^2, S^2] = \mathcal{Z}$. This can be shown by considering local coordinates θ, ϕ and Θ, Φ for the two spheres. A typical homotopy class corresponding to the integer N is $\Theta = \theta$, $\Phi = N\phi$. It counts the number of times the S^2 winds around the target S^2. Hence N is known as the winding number (Fig. 1.5).

Similar arguments can be provided for $\Pi_n(S^n) = \mathcal{Z}$. Also we can establish $\Pi_n(S^m) = \{e\}, m > n$. While it is easy to see $\Pi_n(S^1) = \{e\}, n \geq 2$, the other homotopy groups of S^n require further analysis. Some of these homotopy groups with applications to Quantum Physics will be discussed in Chap. 5.

Fig. 1.5 Winding number N

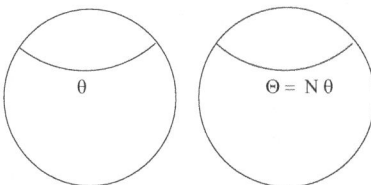

Appendix

- In set S, the equivalence is a binary relation between elements which is (i) reflexive (i.e., the element is equivalent to itself), (ii) symmetric (i.e., $x \approx y \rightarrow y \approx x$), (iii) transitive (i.e., If $x \approx y, y \approx z$, then $x \approx z$.)
- Equivalence relation along with the set gives equivalence class (i.e. Equivalence class $[x]$, $x, y \in [x], x \approx y$)
- A group G is a set of elements $\{g_i\}$ with a product rule '·' such that

  (i) There exists an identity element $e \cdot g_i = g_i \; \forall g_i \in G$
  (ii) $g_i \cdot g_j = g_k$ for any $g_i, g_j, g_k \in G$,
  (iii) associative property $(g_1 \cdot g_2) \cdot g_3 = g_1 \cdot (g_2 \cdot g_3)$ is obeyed and
  (iv) for every element g_i there exists g_i^{-1} such that $g_i \cdot g_i^{-1} = e$.

- subgroup $H \subset G$ with elements of H obeying the above group properties.
- left (right) coset are gH (Hg) where $g \in G$.
- The elements of a group can be equivalently presented by a set of generators and the relations amongst them.
- A free group is a set of generators without any relations amongst them. Then, the group elements are all possible words written in terms of those generators.

Exercises

1.1 Show that the fundamental group of a sphere $\mathbf{S}^2$ with the north pole removed ($\Pi_1(\mathbf{S}^2 - \{N\})$) is trivial. What will be the fundamental group of the sphere $\mathbf{S}^2$ with both north $\{N\}$ and south pole $\{S\}$ removed. Is the topology of $\mathbf{S}^2 - \{N\} - \{S\}$ a disc or a cylinder?

1.2 Give an example of X and subspace A with a map which is retract but not deformation retract.

1.3 Suppose we remove two points x_1, x_2 from $\mathbf{D}^2$. What will be the deformation retract space of $\mathbf{D}^2 - x_1 - x_2$? Show that the fundamental group $\Pi_1(\mathbf{D}^2 - \{x_1\} - \{x_2\})$ is a free group in two variables.

1.4 Explain that, for two touching discs the fundamental homotopy group $\Pi_1(\infty)$ is the same as the the previous problem.

1.5 Show $\Pi_2(\mathbf{S}^2) = \mathcal{Z}$. Extend this to $\Pi_n(\mathbf{S}^n)$.

References

1. H.J. Poincaré, *Les méthodes nouvelles de la mécanique céleste*, vols. 1–3 (Gauthiers-Villars, Paris, 1892, 1893, 1899) (English translation edited by D. Goroff, published by the American Institute of Physics, New York, 1993)
2. G.D. Birkhoff, *Dynamical Systems* (reprinted with an introduction by J. Moser and a preface by M. Morse, 1966) (American Mathematical Society, Providence, 1927)
3. M. Nahakara, *Geometry, Topology and Physics* (Institute of Physics Publishing, London, 2003)
4. C. Nash, S. Sen, *Topology and Geometry for Physicists* (Academic Press, New York, 1983)

Differentiable Manifolds and Geometry 2

In this chapter, we will provide an introduction to smooth manifolds since most of the dynamical system evolutions are described on such spaces.

Topology brings out the continuous structure in spaces. But smooth differentiability is an additional requirement. For example a triangle, square and circle are topologically the same. The cube and the sphere are homeomorphic to each other. But the differentiability of the topological space will face obstructions near the corners. Differentiability will also have continuity and topological invariants can be constructed using smooth functions on such spaces. Hence the need to study smooth differentiable topological spaces which are also known as smooth manifolds.

These smooth manifolds are described in localised small regions by conventional Euclidean geometry and a short introduction to Euclidean geometry is the requisite for developing further.

2.1 Euclidean Geometry, Metric Spaces, Smooth Manifolds

Euclidean geometry is the study of plane, solid figures on the basis of axioms provided by Euclid. It is nothing but plane, three-dimensional geometry and analytical geometry taught in schools. In the classic book 'Elements by Euclid', the only tools required are rulers and the compass. Basic axioms are:

1. For any two points there exists a unique line.
2. A circle can be constructed if a point and a length (radius) are provided.
3. All right angles are equal.
4. For a line L and a point p not on L there exists a line through p not meeting L and this line is unique. This is known as a parallel postulate.
5. The shortest distance between two points x_i, y_i, $i = 1, 2, 3$ is a straight line. The length is given by $s = \sqrt{\sum_i (x_i - y_i)^2}$.

All the known results for Euclidean geometry will follow from the above stated axioms.

2.2 Differentiable Manifolds

A differentiable manifold M is a topological space with a global differential structure. For this we need some background definitions [1] for M to be a topological space.

1. A chart (U, Φ) is an open subset U of M and a homeomorphism Φ from U to an open subset of $\mathbb{R}^n$.

$$\Phi : U \to \mathbb{R}^n \tag{2.1}$$

  This gives a differential structure on M.
2. An atlas is a collection of such charts $\{(U_i, \Phi_i)\}$ on M for which the transition from one chart (U_1, Φ_1) to another (U_2, Φ_2) is made smooth by the requirement $\Phi_2 \cdot \Phi_1^{-1}$ differentiable for elements belonging to $U_1 \cap U_2$.
3. Given the atlas and the differentiable mappings, the differential structure extends to the whole space M.
4. We obtain a smooth differentiable manifold by patching together the above collection of charts and atlases.

Differentiable manifolds can be given locally an Euclidean geometry of $\mathbb{R}^n$ in a small region around any point. The dimension of the manifold is the same as that of the Euclidean space required in a local region. This gives Euclidean coordinates in any local patch (open set).

If two patches overlap then two sets of coordinates will be available for the points in the overlapping regions. But one can go from one coordinate to another by a non-singular transformation.

Definition 2.1 A topological space is a Hausdorff space M if for any two points $x, y \in M$, there are a pair of open sets $\mathcal{O}_x, \mathcal{O}_y$ such that $\mathcal{O}_x \cap \mathcal{O}_y = \phi$ (the null set), no matter how close x and y are.

We also need to define the distance between two points. A topological space endowed with a distance measure is called *a metric space*. Metric is a function

$$d : M \times M \to \mathbb{R}.$$

There exists a shortest-distance function $d(x, y)$ for any pair of points $x, y \in M$ such that:

Definition 2.2

$$d(x,y) \geq 0, \ d(x,y) = 0 \implies x = y; \ d(x,z) \leq d(x,y) + d(y,z)$$

For example in $\mathbb{R}^n$, $d(x,y) = \sqrt{\sum_{i=1}^{n}(x_i - y_i)^2}$. In the Riemannian manifold for two points which are infinitesimally close to each other, we have:

$$d(x,y)^2 = \sum_{i,j}^{n} g_{ij}(x) \, dx^i \, dx^j \tag{2.2}$$

where $y^i = x^i + dx^i$. We will call g_{ij} as the metric tensor components. For arbitrary points we need further information. We connect the points along a curve $\mathcal{C}$ parametrized by $x^i(t)$, then the distance along the curve is

$$d_\mathcal{C}(x,y) = \int_\mathcal{C} \sqrt{g_{ij} \frac{dx^i}{dt} \frac{dx^j}{dt}} \, dt \tag{2.3}$$

The geodesic distance between any two arbitrary points is defined as the least value of $d_\mathcal{C}(x,y)$ when we vary the curves. That curve which provides the least distance is also known as geodesic.

2.3 Differential Calculus on Manifolds

We are familiar with calculus on $\mathbb{R}^n$. Hence for any differentiable manifold M covered by local patches(charts and atlas), we will introduce differential calculus. For an introduction to differential calculus on manifolds see Refs. [2, 3]. Let $x^i[(p(t)]$ be the coordinates of a point p along a curve $\mathcal{C}$ parametrising t in a local patch.

We define 'velocity vector' or tangent vector as

$$\frac{dx^i[p(t)]}{dt}.$$

Consider $f(p)$ a real-valued function (for example it can be temperature along a curve) and find how it changes along the curve. It is

$$\frac{df}{dt} = \sum \frac{\partial f}{\partial x^i} \frac{dx^i}{dt} \tag{2.4}$$

We can write this as

$$\frac{df}{dt} = X f \text{ where } X = \sum_{n}^{n} a^i \frac{\partial}{\partial x^i}.$$

Following Einstein's convention we will drop the summation sign hereafter and assume repeated indices are summed over. Here $a^i = \frac{dx^i[p(t)]}{dt}$, are the components of tangent vector. In general any linear combination $a^i \frac{\partial}{\partial x^i}$ is tangent to some curve passing through $p(t)$.

The span of all such tangents at a point is called the tangent space $T_p(M)$ at p and $\frac{\partial}{\partial x^i}$ form the basis of this vector space. These are also known as *contravariant vectors*. The collection of all such tangent spaces for all open sets can be given a differentiable structure and is known as the *tangent bundle*. It has a dimension $2n$ for the n-dimensional manifold M.

What are covariant vectors? They correspond to the dual space denoted by $T_p^*(M)$. Following the earlier definition, the collection of all the cotangent spaces is known as the *cotangent bundle*.

A simple example of a covariant vector is the so-called one form obtained through the basis dx^i. An arbitrary function like for example density or temperature on the manifold will change from point to point. If we displace x^i by $x^i + dx^i$ (with infinitesimal dx^i) then the function will change to $f(x^i + dx^i) = f(x^i) + df(x^i)$ where

$$df = \frac{\partial f}{\partial x^i} dx^i. \tag{2.5}$$

The basis for the vector space $T_p^*(M)$ is given by dx^i and an arbitrary element can be written as $a_i \, dx^i$. A one-form acts as a covariant vector. An element of $T_p^*(M)$ should act on an element $X \in T_p(M)$ to give a real (or complex number).

$$(df, X) = X[f] = X^i \frac{\partial}{\partial x_i} f,$$

where the basis obeys

$$\left(dx^i, \frac{\partial}{\partial x^j}\right) = \delta_j^i.$$

Consider an arbitrary one-form $\omega = a_i \, dx^i$ and an arbitrary contravariant vector $X = b^i \frac{\partial}{\partial x^i}$. Using the above relation, it follows that

$$(\omega, X) = a_i \, b^i \tag{2.6}$$

Definition 2.3 (Exterior Product) Given two one-forms ω, μ (which are also covariant vectors) we can obtain a two-form through the exterior product:

$$\omega \wedge \mu = \frac{1}{2} (\omega_i \, \mu_j - \omega_j \, \mu_i) \, dx^i \, dx^j. \tag{2.7}$$

2.3 Differential Calculus on Manifolds

Geometrically the product $dx^i dx^j$ provides an elemental area of parallelogram constructed out of the two elementary lengths dx^i. Now we define the wedge product between the elements dx^i, dx^j as

$$dx^i \wedge dx^j = - dx^j \wedge dx^i .$$

Using this, the exterior product can be written as

$$\omega \wedge \mu = \omega_i \mu_j dx^i \wedge dx^j \tag{2.8}$$

Wedge product is equivalent to taking cross-product in three dimensions. It provides the basis for antisymmetric tensors. This can be extended to completely antisymmetric higher-rank tensors. The highest rank we can obtain is provided by the dimension of the manifold.

The space of antisymmetric second rank tensors denoted by $\Lambda^2 T_p^*(M)$. The space of vectors is called $\Lambda^1 T_p^*(M)$ and the space of scalars is $\Lambda^0 T_p^*(M)$. The dimension of the vector space $\Lambda^2 T_p^*(M)$ can be easily seen to be ${}^nC_2$, where n is the dimension of the manifold M. We can extend the above to a space of r-forms as

$$\alpha = \alpha_{i_1 i_2 \ldots i_r} dx^{i_1} \wedge dx^{i_2} \wedge \ldots . dx^{i_r} \tag{2.9}$$

This provides a completely antisymmetric tensor $\alpha_{i_1 i_2 \ldots i_r}$. The *rank* of the form is also known as *degree* of the form.

Definition 2.4 Given $\alpha \in \Lambda^r T_p^*(M), \beta \in \Lambda^s T_p^*(M)$ we can define the exterior product of the two forms leading to $(r+s)$ form as:

$$\alpha \wedge \beta = (-1)^{rs} \beta \wedge \alpha, \quad \alpha \wedge (\beta \wedge \gamma) = (\alpha \wedge \beta) \wedge \gamma \tag{2.10}$$

The exterior product is associative but not commutative in general.

From antisymmetric tensors at a point in $T_p^*(M)$ we can move on to tensor fields known as forms on M by allowing them to vary continuously over the whole manifold.

If ω is an r-form, it can be represented by

$$\omega = \omega_{i_1 i_2 \ldots i_r}(x) dx^{i_1} dx^{i_2} \ldots dx^{i_r} \tag{2.11}$$

Note We have dropped the wedge symbol for convenience. It is taken as understood and we will indicate them whenever clarity is needed. The space of such r forms on the manifold M is labelled by $\Omega^r(M)$.

If ω is an r form $\in \Omega^r(M)$ then $dim(\Omega^r(M)) = {}^nC_r$. Hence if $r > n$ then $\omega = 0$. In local coordinates, we can write

$$\omega(x) = f_{i_1 i_2 \ldots i_r}(x)\, dx^{i_1} dx^{i_2} \ldots dx^{i_r} \tag{2.12}$$

We need to differentiate the fields and express the changes in a covariant way. This leads us to the definition of exterior derivative.

Definition 2.5 (Exterior Derivative) Given $\omega \in \Omega^r(M)$, we define the exterior derivative as $d\omega \in \Omega^{r+1}(M)$:

$$d\omega = \frac{\partial f_{i_1 i_2 \ldots i_r}}{\partial x^{i_{r+1}}} dx^{i_{r+1}} dx^{i_1} dx^{i_2} \ldots dx^{i_r} \tag{2.13}$$

In effect the exterior derivative maps an r-form to an $r+1$- form:

$$d : \Omega^r(M) \longrightarrow \Omega^{r+1}(M) \tag{2.14}$$

We should remember the dimensions of the space r forms and $r+1$ forms are different in general. We then have the following identities.

- $d^2 \omega = 0$.
  Proof

$$d^2 w = \frac{\partial^2 f_{i_1 i_2 \ldots i_r}}{\partial x^{i_{r+2}} \partial x^{i_{r+1}}} dx^{i_{r+2}} dx^{i_{r+1}} dx^{i_1} \ldots dx^{i_r} = 0 \tag{2.15}$$

The second derivative in the above equation is symmetric in $x^{i_{r+1}}$ and $x^{i_{r+2}}$ whereas the wedge product $dx^{i_{r+1}} \wedge dx^{i_{r+2}}$ is antisymmetric and the summation leads to the result.
- $d(\alpha + \beta) = d\alpha + d\beta$
- $d(\alpha \wedge \beta) = d\alpha \wedge \beta + (-1)^{\deg \alpha} \alpha \wedge d\beta$ where $\deg \alpha$ denotes degree or rank of the form α.
- If $f \in \Omega^0(M)$ is a function (zero form) then $df = \vec{\nabla} f \cdot \vec{dr}$
- In 3 dimensions, if ω is one form $d\omega$ is related to 'curl' $\vec{\nabla} \times \vec{\omega}$ defined as usual in vector calculus textbooks. If ω is a two-form then $d\omega$ is related to the divergence.

Since ${}^nC_r = {}^nC_{n-r}$, we have the 'dimension of space of r-forms same as the dimension of the space '$n - r$'-forms'. That brings in the question of mapping $\Omega^r(M)$ to $\Omega^{n-r}(M)$. We can define dual operator which maps $\Omega^r(M) \longrightarrow \Omega^{n-r}(M)$ in local coordinates. Note the $dim(\Omega^r(M)) = dim(\Omega^{n-r}(M)$. Let $\alpha \in \Omega^r(M)$. We write in local coordinates

$$\alpha = \alpha_{i_1 i_2 \ldots i_r}\, dx^{i_1} dx^{i_2} \ldots dx^{i_r}$$

We define ${}^*\alpha \in \Omega^{n-r}$ as:

2.3 Differential Calculus on Manifolds

Definition 2.6

$$*\alpha = \frac{1}{(n-r)!} \alpha_{i_1 i_2 \ldots i_r} \epsilon^{i_1 i_2 \ldots i_r i_{r+1} \ldots i_n} dx^{i_{r+1}} dx^{i_{r+2}} \ldots dx^{i_n}$$

This is known as the Hodge star operator. Note we have used the Levi-Civita symbol $\epsilon^{i_1 i_2 \ldots i_r i_{r+1} \ldots i_n}$ which is completely antisymmetric in all the indices and $\epsilon^{123\ldots n} = 1$. Also $\epsilon^{i_1 i_2 \ldots i_r i_{r+1} \ldots i_n}$ which is obtained by an even permutation of $\{12..n\}$ is equal to 1. Odd permutation gives -1. Now we will provide a simple physical application of the above description of forms. It is important to appreciate the above formalism is entirely coordinate free and we need not work with specific coordinates.

Examples

1. We have in electromagnetic theory $\vec{A}$ the vector potential and ϕ the scalar potential. Both of them together transform as a four-vector $A_\mu = (\phi, \vec{A})$ in Minkowski space. We can write what is known in the literature as the connection one form A as :

$$A = A_\mu \, dx^\mu \tag{2.16}$$

Here we have used the unit where the velocity of light $c = 1$. This will give as immediately the field strength or curvature two-form[1]

$$F = F_{\mu\nu} \, dx^\mu \, dx^\nu = dA \tag{2.17}$$

We also have the identity

$$dF = d^2 A = 0 \implies \partial_\mu F_{\nu\lambda} + \partial_\nu F_{\lambda\mu} + \partial_\lambda F_{\mu\nu} = 0 \tag{2.18}$$

This is known as Bianchi identity in electromagnetism. The electric field $\vec{E}$ and magnetic field $\vec{B}$ are given by

$$E_i = -F_{0i} = \partial^0 A^i - \partial^i A^0 \; ; \; B_i = \frac{1}{2} \epsilon_{ijk} F^{jk} = (\vec{\nabla} \times \vec{A})_i.$$

This gives the equation $dF = d(dA) = 0$ as:

$$\nabla \cdot \vec{B} = 0, \quad \nabla \times \vec{E} = -\frac{\partial \vec{B}}{\partial t}. \tag{2.19}$$

---

[1] $\wedge$ symbol is suppressed.

These two of Maxwell's equations follow from the definition of field strength $\vec{E}$ and $\vec{B}$ through 4-potential A_μ. Given a two-form $F = dA$ we can get its dual two-form as $^*F = ^*dA$. We can show that

$$^*\left(d^*F\right) = \partial_\mu F^\mu_\lambda \, dx^\lambda . \tag{2.20}$$

The above one-form equated to zero

$$^*\left(d^*F\right) = \partial_\mu F^\mu_\lambda \, dx^\lambda = 0$$

gives the two Maxwell's equations in free space without any charges or currents:

$$\nabla \cdot \vec{E} = 0, \quad \nabla \times \vec{B} = \frac{\partial \vec{E}}{\partial t} . \tag{2.21}$$

Hence, we can write Maxwell equations concisely as:

$$F = dA, \quad dF = 0, \quad ^*d^*F = 0 \tag{2.22}$$

We can also write the last equation as $d^*F = 0$.

If there are sources for the electric and magnetic fields, that is charge and current densities, we can define four-current density as $j_\mu = (\rho, \vec{j})$ and define current one form as $j = j_\mu \, dx^\mu$. Then Maxwell's equations with sources are:

$$F = dA, \quad dF = 0, \quad ^*d^*F = j \tag{2.23}$$

Since Maxwell's equations are expressed in terms of only $F = dA$ they have a new symmetry known as $A \to A' = A + d\lambda$ where λ is a zero form (scalar function). We write this as $A' = A + e^{-i\lambda} d\, e^{i\lambda}$. Here $e^{i\lambda}$ can be taken as an element of the simplest unitary group denoted as $U(1)$. In fact, the electromagnetic theory reflects $U(1)$ gauge symmetry which is an abelian symmetry group. This can be generalised to other nonabelian group symmetry.

2. We can generalize the above analysis to define two-form 'potential'. This is known as Kalb-Ramond [4] field in literature.

$$B = B_{\mu\nu} dx^\mu \, dx^\nu \tag{2.24}$$

The corresponding 'field strength' $H = dB$ is a three-form field. That is:

$$H = H_{\mu\nu\lambda} dx^\mu \, dx^\nu \, dx^\lambda \tag{2.25}$$

The corresponding field equations are

$$dH = 0, \quad ^*d^*H = 0. \tag{2.26}$$

2.3 Differential Calculus on Manifolds

These are equations for the Kalb-Ramond fields. One can establish it describes a massless scalar field.

3. The vector potential A_μ in $2+1$ dimensions introduces novelties. This is because $F = dA$ is dual to another one- form in three dimensions. Due to this property, one can write an equation of the form:

$$d^* F = m F + {}^*j \qquad (2.27)$$

Here the parameter m has the dimensions of mass. This introduces mass to the particle in a novel gauge invariant way. Also, this equation of motion plays an important role in describing quantum hall systems in condensed matter physics. When the left- hand side is absent, the equation results from an action known as Chern Simons term. This leads to interesting topological properties which we will elaborate in a later chapters on Chern-Simons theory.

4. Further generalisation of the vector potential to Lie algebra valued one form defined as:

$$A = A_\mu^a T^a dx^\mu \qquad (2.28)$$

where T^a are the generators of the nonabelian group like higher rank unitary groups $SU(N)$ lead us to the Yang-Mills equations. These are generalisations of Maxwell's theory. The new gauge transformation is $A' = g^{-1} A g + g^{-1} dg$ where the $SU(N)$ group element $g = e^{i\lambda^a T^a}$ generalises the earlier $U(1)$ symmetry. The Lie algebra valued two-forms is obtained in such a way it automatically transforms $F \to F' = g^{-1} F g$. The definition of the two-form field strength is given by

$$F = dA + A \wedge A \qquad (2.29)$$

We can show it reflects the covariant transformation property. i.e., $F' = g^{-1} F g$. Yang-Mills equation without sources/currents now becomes:

$$D {}^*F = 0 \qquad (2.30)$$

where $D = d + ig A = \partial_\mu + ig A_\mu^a T^a$ is known as the covariant derivative. We will elaborate on these with respect to topological information in the later chapters.

2.3.1 Closed and Exact Forms

Now we have developed an important framework which provides the basis for new topological invariants. We will now define a closed and exact form.

Definition 2.7

1. ω is closed if $d\omega = 0$.
2. ω is an exact form if there exists a form α such that $\omega = d\alpha$.

Now we define $\mathbf{Z}^p$ as the space of all closed p forms:

Definition 2.8 If $\omega^{(p)} \in \mathbf{Z}^p$, then $d\omega^{(p)} = 0$.

Any two closed p forms can be added with linear coefficients to get another closed p form.

Again we define $\mathbf{B}^p$ as the space of all exact p forms:

Definition 2.9 If $\omega^{(p)} \in \mathbf{B}^p$, then $\omega^{(p)} = du^{(p-1)}$.

Two exact p forms can again be added to give another exact form.

We call two p-forms equivalent if they differ by an exact form i.e., $\omega_1^{(p)} \approx \omega_2^{(p)}$ if

$$\omega_1^{(p)} = \omega_2^{(p)} + du^{(p-1)} \tag{2.31}$$

where $u^{(p-1)}$ is a $p-1$ form.

We can define topological invariants using the set of closed forms which are not expressible as exterior derivatives of an exact form. These are written as the coset space: $\frac{\mathbf{Z}_p}{\mathbf{B}_p}$.

To understand these invariants we need to study integration on differentiable manifolds. That will bring us to new topological invariants known as homology and cohomology [5, 6].

2.4 Homology

In the previous sections, we developed the idea of topological invariants of differentiable manifolds. However, the smooth manifolds are not needed to study the topological properties. We simplify the spaces by considering them to be made up of various patches stitched together. For example a circle $\mathbf{S}^1$ can be considered as homeomorphic to a triangle. A two-dimensional disc $\mathbf{D}^2$ has the same topological invariants as a triangle with interior included. A sphere $\mathbf{S}^2$ has the same invariants as a tetrahedron or a cube. A three-dimensional ball $\mathbf{B}^3$ is a tetrahedron or cube with an included interior. We generalize these notions by defining what is known as a 'simplicial complex'.

For this purpose, we cover the manifold with smaller units called polyhedra which are subsets of $\mathbb{R}^n$. The process of covering the manifold is called triangulation. This requires first a definition of a simplex.

2.4 Homology

A set of points x_i, $i = 1, 2, \ldots m+1$ are independent if the vectors $x_i - x_1$, $\forall i \neq 1$ are linearly independent.

An **m-simplex** σ^m in $\mathbb{R}^n$ is defined as:

$$x = \sum_1^{m+1} \lambda_i x_i, \quad \lambda_i \geq 0, \quad \sum \lambda_i = 1 \quad (2.32)$$

and x_i are independent. It is denoted by

$$\sigma^m = [x_1, x_2, \ldots x_{m+1}]$$

and the points are the vertices of the simplex. We can regard x as the center of mass for the system with masses λ_i being placed at vertices x_i, $\forall i$.

The **Faces** of a simplex are the set $\sum_{i \neq j} \lambda_i x_i$ with $\lambda_j = 0$. The above is called the jth face of the simplex opposite to the vertex x_j

We now provide some simple examples:

Examples

- zero simplex is a point x_1.
- A 1-simplex is closed interval $[x_1, x_2]$ such that $x = \lambda_1 x_1 + (1 - \lambda_1) x_2$.
- A 2-simplex is a given by a triangle with all its interior points including the boundary. i.e., $x = \lambda_1 x_1 + \lambda_2 x_2 + (1 - \lambda_1 - \lambda_2) x_3$.

We have drawn 1, 2, 3-simplices in Fig. 2.1.

Fig. 2.1 1,2,3-Simplices

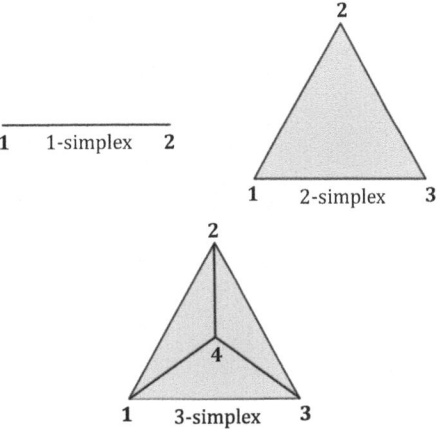

If $\lambda_1 = 0$, then we get the edge $[x_2, x_3]$ and similarly two more edges. It is obvious we cannot get a 3-simplex in $\mathbb{R}^2$. We need at least 3 dimensions. In $\mathbb{R}^3$ a tetrahedron $[x_1, x_2, x_3, x_4]$ is a 3 simplex. Again we get the face $[x_2, x_3, x_4]$ by imposing $\lambda_1 = 0$. We have four such faces opposite to each vertex.

We now define an open simplex as the interior of any closed simplex.

$$\sigma^m = \left\{ x = \sum_1^{m+1} \lambda_i x_i, \ \sum \lambda_i = 1, \ \lambda_i > 0 \right\} \qquad (2.33)$$

Armed with the above definitions we can now define a **Simplicial complex**. It is the collection of closed simplices, along with their faces. The dimension of the complex is the maximum dimension of the simplices. We can also perform deformation to a simplex as shown in Fig. 2.2.

Polyhedron This is the union of all members of the simplicial complex with Euclidean subspace topology. In certain literature the term 'polytope' is used. For example, the tetrahedron (Fig. 2.3) consists of four 0-simplex, six 1-simplex (edges), four two simplex and one three simplex.

We will now provide a simple example of the triangulation of $\mathbf{S}^1$ and $\mathbf{D}^2$ (disc). The same will be useful in explaining the topological invariants of these.

1. Triangulation of $\mathbf{S}^1$. This polyhedron K consists of (see Fig. 2.4)

$$K = [1], [2], [3], [1, 2], [1, 3], [2, 3]$$

2. Triangulation of $\mathbf{D}^2$ is $K = [1], [2], [3], [1, 2], [1, 3], [2, 3], [1, 2, 3]$ (see Fig. 2.4)
3. Cylinder is given by the triangulation as shown in Fig. 2.5
4. Torus is given by the triangulation as depicted in Fig. 2.6

Fig. 2.2 Deformation to a simplex

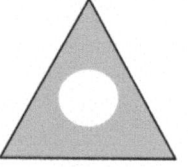 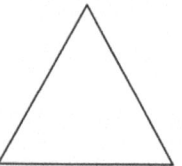

Fig. 2.3 Polyhedron

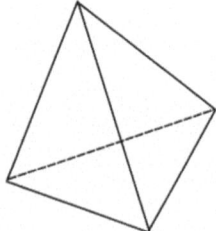

Fig. 2.4 Triangulation of circle and disc

Fig. 2.5 Triangulation of cylinder

Fig. 2.6 Triangulation of torus

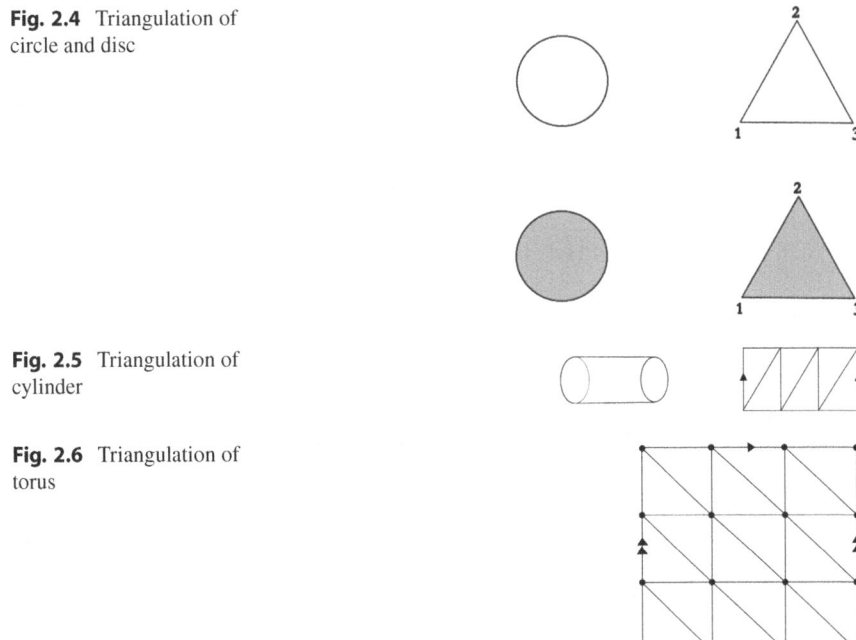

As pointed out earlier, triangulation is not unique. But invariants will be the same for any triangulation.

Path Connected Complex We can ask the question whether we can go from one vertex to another one along edges. This leads to the definition of path connected space. A simplicial complex is path connected if for every pair of vertices u, v there exists a sequence of vertices such that $v_i v_{i+1}$ is a one simplex. That is there is a path between any two vertices. For example $[x_1, x_2, x_3]$ is path connected. That is any two vertices are connected.

2.5 Simplicial Complex and the Fundamental Group

Having defined the simplicial complex which is homeomorphic to a manifold of interest we can obtain the fundamental group of the manifold through defining generators and the relations.

Consider K a path-connected polyhedron. We first obtain the one-dimensional sub-polyhedron which is contractible to a point and contains all the vertices. We associate a generator g_{ij} for each 1-simplex $[i, j]$. We introduce a relation $g_{ij} g_{jk} g_{ki}^{-1} = 1$ for each 2-simplex $[i, j, k]$ containing this simplex $[i, j]$. Then the group generated by these generators with these conditions is the fundamental group of the simplicial complex which in turn is homeomorphic to the manifold.

The important result which follows is that 1 and 2 simplices alone determine the fundamental group.

We will now provide a few examples using the above construction to obtain the fundamental group.

1. $\pi_1(\mathbf{S}^1)$: The polyhedron corresponding to S^1 can be given as

$$[1], [2], [3], [1, 2], [2, 3], [1, 3].$$

The sub-polyhedron containing all the vertices and contractible to a point is $[1, 3], [2, 3]$. If we associate a generator for each one simplex and provide the relation with those with homotopy type of a point as 1 we get $g_{12} = g$, $g_{13} = g_{23} = 1$ we get our fundamental group is generated by one element g. That proves $\pi_1(\mathbf{S}^1) = \mathcal{Z}$.

2. $\pi_1(\mathbf{D}^2)$: Here we have the same polyhedron of the previous example along with the addition of the 2-simplex $[1, 2, 3]$. Now the new relation $g_{12}g_{23}g_{13}^{-1} = 1$. But we already have $g_{13} = g_{23} = 1$ being part of contractible one simplex. That makes $g_{12} = 1$. Hence the $\pi_1(\mathbf{D}^2) = 1$.

3. $\pi_1(Cylinder = [1, 2] \otimes \mathbf{S}^1)$: The triangulation has 6 vertices, 12 edges, 6 faces (See Fig. 2.7). The simplices are:

   - 0-simplex: $\sigma^0 = [1], [2], [3], [4], [5], [6]$,
   - 1-simplex: $\sigma^1 = [1, 2], [2, 3], [3, 1], [4, 5], [5, 6], [6, 4], [1, 4], [1, 5], [2, 5], [2, 6], [3, 6], [3, 4]$
   - 2-simplex: $\sigma^2 = [1, 2, 5], [1, 4, 5], [2, 5, 6], [2, 3, 6], [1, 3, 4], [3, 4, 6]$.

   The contractible 1-simplex is $[2, 3], [3, 6], [6, 4], [4, 5], [5, 1]$. These can be taken as 1. Other 1-simplices which are part of 2 simplices can be simplified using the relation. The independent generator is only one and that is g_{12} and hence $\pi_1([1, 2] \otimes \mathbf{S}^1) = \mathcal{Z}$.

4. $\pi_1(\mathbf{S}^2)$: The polyhedron consists of

$$[1], [2], [3], [4], \quad [1, 2], [1, 3], [1, 4], [2, 3], [2, 4], [3, 4],$$
$$[1, 2, 3], [1, 3, 4], [2, 3, 4], [1, 2, 4]$$

   The 1-simplex contractible to a point is $[1, 2], [2, 3], [3, 4]$. They can be taken as 1. Then we have generators g_{13}, g_{24}, g_{14}. Now the 2-simplex $[1, 2, 3]$ will give $g_{12}g_{23}g_{13}^{-1} = 1$ But we already have $g_{12} = g_{23} = 1$. Then we obtain $g_{13} = 1$. Similarly we can prove $g_{24} = g_{14} = 1$. Hence we get $\pi_1(\mathbf{S}^2) = 1$.

5. $\pi_1(\mathbf{T}^2)$: The triangulation (Fig. 2.8) with 7 vertices (0-simplices), 14 edges (1-simplices), 14 (2-simplices). The contractible path is $[1, 2], [2, 3], [3, 4], [4, 5], [5, 6], [6, 7]$. The total number of generators is $^7C_2 = 21$. Hence the remaining generators are 15. They satisfy 14 relations coming from 14 two simplices. Left over are two generators which commute among themselves giving $\pi_1(\mathbf{T}^2) = \mathcal{Z} \oplus \mathcal{Z}$.

2.5 Simplicial Complex and the Fundamental Group

Fig. 2.7 Cylinder triangulation

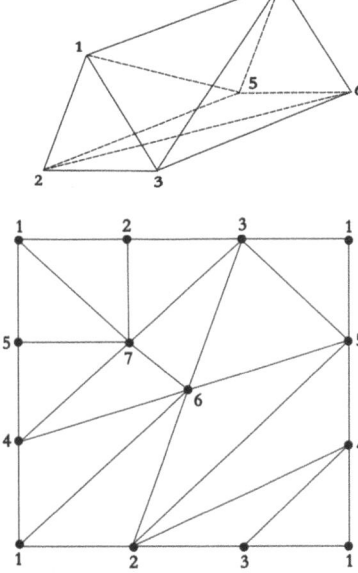

Fig. 2.8 Torus complex with 7 vertices

Orientation Choose an ordering of the vertices of a simplex. Any even permutation of that ordering is considered as positive orientation. Example:

$$\sigma^2 = [x_1, x_2, x_3] = [x_2, x_3, x_1] = [x_3, x_1, x_2].$$

However, the odd permutation referred to as negative orientation: $-\sigma^2 = [x_3, x_2, x_1] = [x_1, x_3, x_2] = [x_2, x_1, x_3]$

To develop the notion of homological invariants we need to introduce a p-chain.

Definition 2.10 (p-Chain) $\mathcal{C}_p$ is simply a formal sum of p-simplices with integer coefficients.

$$\mathcal{C}_p = \sum n_i \sigma_i^{(p)}. \tag{2.34}$$

We can consider them to be an element of the abelian group of integers $\sum_\oplus \mathcal{Z}$

Definition 2.11 (Boundary Operator) Boundary Operator is defined as ∂_p: acting on a p chain gives $p - 1$ chain.

$$\partial_p : \mathcal{C}_p \longrightarrow \mathcal{C}_{p-1} \tag{2.35}$$

It acts linearly

$$\partial_p \sum_i f_i \sigma_i^{(p)} = \sum_i f_i \partial_p \sigma_i^{(p)}$$

It is given by:

$$\partial_p \sigma^{(p)} = \partial_p [x_0, x_1, x_2, \ldots, x_p] = \sum_0^p (-1)^i [x_0, x_1 \ldots \hat{x}_i, \ldots x_p] \quad (2.36)$$

where $[x_0, x_1 \ldots \hat{x}_i, \ldots x_p]$ represents $\sigma^{(p-1)}$ obtained from $\sigma^{(p)}$ by omitting the vertex x_i.

Example

$$\partial_2 [x_0, x_1, x_2] = [x_0, x_1] - [x_0, x_2] + [x_1, x_2]$$

It is easy to prove the following important result:

Theorem 2.1

$$\partial_{p-1} \cdot \partial_p = 0 \quad (2.37)$$

That is, the boundary of a boundary is null. Boundary operator formally provides the boundary of a manifold.

Example

$$\partial_1 \cdot \partial_2 [x_0, x_1, x_2] = \partial_1 ([x_0, x_1] - [x_0, x_2] + [x_1, x_2])$$
$$= [x_0] - [x_1] - [x_0] + [x_2] + [x_1] - [x_2] = 0$$

We can identify number of 'holes' in a space using p-chains. For that we need **p-cycle:**

Definition 2.12 A chain $z_p \in \mathcal{C}_p$ is a p-cycle if $\partial_p z_p = 0$. That is, it does not have a boundary(like a loop which is homeomorphic to a circle).

Hence these cycles are mapped by boundary operator to the identity element of $\mathcal{C}_{p-1}$. That is, they are Kernel of the maps $\mathcal{C}_p \longrightarrow \mathcal{C}_{p-1}$. That is, the inverse image of the identity element in $\mathcal{C}_{p-1}$. This set of elements $z_p \in \mathcal{C}_p$ must be a subgroup of the abelian group $\mathcal{C}_p$. Let us call them as Z_p.

2.5 Simplicial Complex and the Fundamental Group

Definition 2.13 A chain $b_p \in C_p$ is called **p-boundary** if there exists a $p+1$ chain C_{p+1} such that

$$\partial_{p+1} C_{p+1} = b_p$$

That is, the set of elements is the image of the map $C_{p+1} \longrightarrow C_p$. Basically, we have a map from the abelian group C_{p+1} to another abelian group C_p. We are looking at the image of this map. Naturally, this is also a subgroup of C_p. Call this as B_p. Since $\partial_p b_p = 0$, automatically B_p must be a subgroup of Z_p.

If we remove the elements of B_p in Z_p we get the definition of p-dimensional holes. What is left out after removing those which are boundaries are genuine p-dimensional holes. We associate integers $\mathcal{Z}$ to each of these holes. These define the homology group of the polyhedron K which is the simplicial complex obtained by triangulating the smooth manifold. The manifold and the polyhedron are homeomorphic to each other.

Definition 2.14 Hence we define homology group $H_p(K)$ by quotienting Z_p by B_p as:

$$H_p(K) = \frac{Z_p}{B_p}$$

An element of H_p is in the equivalent class $[z_p]$. That is, $z_p^{(1)}$ and $z_p^{(2)}$ are equivalent if $z_p^{(1)} - z_p^{(2)} \in B_p$. Homology groups are topological invariants of the smooth manifold from which the simplicial complex is obtained.

For example $H_1(K)$ consists of those loops which are not boundary loops of a cylinder!. Further, it is easy to see on $\mathbf{S}^1$ that $H_1(S^1) = \mathcal{Z}$. But on $\mathbf{T}^2$, we have two non-trivial 1-cycles known as meridian and longitude. Hence we have $H_1(\mathbf{T}^2) = \mathcal{Z} + \mathcal{Z}$.

A topological space which is homeomorphic to a polyhedron is said to be triangulable. The polyhedron is the triangulation of the space. It is not unique. It is like coarse-graining the space and can be made more and more fine.

Different triangulations can give rise to different p-cycle groups Z_p, p-boundary groups B_p. But the homology groups H_p are always the same for any triangulation.

If the space is contractible (like $\mathbb{R}^n$ without any holes), then $H_p = 0$, $\forall\, p \neq 0$ and $H_0 = \mathcal{Z}$.

Examples

$$K = \{\sigma^2 = [x_0, x_1, x_2], [x_0, x_1], [x_1, x_2], [x_2, x_0], [x_0], [x_1], [x_2]\}$$

we have $H_p(K) = 0$, $p > 2$, $H_0(K) = \mathcal{Z}$, $H_1(K) = 0$, $H_2(K) = 0$.
For a circle the triangulation:

$$K = [x_0, x_1], [x_1, x_2], [x_2, x_0], [x_0], [x_1], [x_2],$$

the groups are: $H_0(K) = \mathcal{Z}$, $H_1(K) = \mathcal{Z}$

2.6 Cohomology

In this section, we will study the groups dual to the homology groups. To do this we need to go through integration of differential forms. For this, we need a 'volume' element or 'measure' on the manifold.

The dual to the homology is indicated by cohomology groups. We can use both homology and cohomology to characterise the invariants associated with the manifolds. Cohomology is more useful and also powerful tool for applications to physical contexts.

Consider an n-form ω in a patch U_α.

$$\omega = dx^1 \wedge dx^2 \wedge \ldots dx^n$$

This n-form is a candidate for the 'volume'. Let us shift to overlapping patch U_β with local coordinates y_i Then

$$\omega = \frac{\partial x^1}{\partial y^{i_1}} dy^{i_1} \wedge \frac{\partial x^2}{\partial y^{i_2}} dy^{i_2} \ldots \frac{\partial x^n}{\partial y^{i_n}} dy^{i_n}$$

$$= \left| \frac{\partial x^i}{\partial y^j} \right| dy^1 \wedge dy^2 \ldots \wedge dy^n$$

This determinant which appears is the familiar Jacobian. Now if we want the integral over U_α:

$$\int_{U_\alpha} f\,\omega = \int f(x_1, x_2, ...x_n) dx_1\, dx_2\, \ldots dx_n$$

This integral can be continued over all the overlapping patches using the Jacobian. The answer remains the same even if different patches were chosen.

2.6 Cohomology

For a Riemannian manifold

$$\omega = \sqrt{g}\, dx^1 \wedge dx^2 \ldots \wedge dx^n$$

is the useful volume element. This makes ω independent of coordinates transformations. We have introduced $\sqrt{g}$ in the above which results from the definition of metric for a Riemannian metric. We will explain metric tensors and their components in the next chapter.

Theorem 2.2 (Stoke's Theorem) *Consider a one form* $\omega = df$ *and integral along an interval* $[a, b]$. *Then*

$$\int_{M=[a,b]} df = \int_a^b \frac{\partial f}{\partial x}\, dx = f(b) - f(a)$$

Consider a two form: $\omega = dA$ *where A is a one form. Given M as a surface in $\mathbb{R}^n$ we have:*

$$\int_M \omega = \int_M dA = \int_M \partial_i A_j\, dx^i \wedge dx^j = \int \nabla \times A \cdot d\vec{S} = \oint A \cdot d\vec{l}$$

Here A is a vector field in $\mathbb{R}^n$ and $d\vec{S}$ is an area element of the surface and $d\vec{l}$ is a line element of the boundary. This can be written as $\int_M dA = \int_{\partial M} A$, where ∂M is the boundary of M.

Theorem 2.3 (Gauss Theorem) $\int_V \nabla \cdot E\, dV = \int_{\partial V} \vec{E} \cdot d\vec{S}$.

These results suggest generalization to n-dimensional manifold with boundary.

Theorem 2.4 (Theorem)

$$\int_M d\omega = \int_{\partial M} \omega$$

Here M is n-dimensional manifold and ∂M is the boundary and is $n-1$ dimensional manifold.

Recall the definition of a closed form ω: $d\omega = 0$ and the exact form where $\omega = df$.

2.6.1 Integration on a Manifold with Boundary

An oriented patch (chart) on a manifold induces an oriented manifold on the boundary. Since the patch has the geometry of $\mathbb{R}^n$, we need a patch with a boundary. Instead of $\mathbb{R}^n$, we can use $\frac{1}{2}\mathbb{R}^n$, $with$ $x_n \geq 0$. Then, we can see $\partial(\frac{1}{2}\mathbb{R}^n) = \mathbb{R}^{n-1}$. If $\omega = dx^1 \wedge dx^2 \wedge dx^2 \ldots \wedge dx^n$ then

$$i_n \omega = (-1)^n \, dx^1 \wedge dx^2 \ldots \wedge dx^{n-1}$$

Hence

$$\int_M d\omega = \int_{\partial M} i^* \omega.$$

Here $\omega = n-1$ form on M. and

$$i : \partial M \longrightarrow M$$

is inclusion map for any point 'p' on ∂M i.e $i(p)$ is the same point in M.

Let

$$\omega = \sum (-1)^{j-1} h^j dx^1 \, dx^2 \ldots dx^{j-1} \, dx^{j+1} \ldots dx^n$$

where we have omitted $\wedge$ symbols. Then

$$dw = \frac{\partial h^j}{\partial x^j} dx^1 \ldots dx^n$$

From the above it follows if $U_\alpha \cap \partial M = \{\phi\}$ (null set):

$$\int dw = \int_a^a \frac{\partial h^j}{\partial x^j} dx^1 \ldots dx^n = 0$$

If $U_\alpha \cap \partial M \neq \{\phi\}$ then

$$\int d\omega = \int dx^1 \int dx^2 \ldots \int_0^a dx^n \frac{\partial h^n}{\partial x^n}$$
$$= -\int \ldots \int h^n(x^1, x^2 \ldots x^{n-1}, 0) dx^1 dx^2 \ldots dx^{n-1}$$

2.6 Cohomology

Remember on ∂M we have $x^n = 0$

$$i^*\omega = (-1)^{n-1} h^n(x^1, x^2, \ldots x^{n-1}, 0) dx^1 \, dx^2 \ldots dx^{n-1}$$

Hence we have

$$\int_M d\omega = \int_{\partial M} i^*\omega \qquad (2.38)$$

is the generalization of both Stokes and Gauss's theorem in classical electrodynamics. It is generally known as Stokes theorem. If we have a p-form integrated over a p-chain $\mathcal{C}_p$ we get a real number.

$$\omega : \mathcal{C}_p \longrightarrow R$$

We write this as $\langle \omega, \mathcal{C} \rangle = \int_\mathcal{C} \omega$. We can call p-form as p-cochain! Hence Stokes theorem is:

Theorem 2.5

$$\langle \omega, \partial \mathcal{C} \rangle = \langle d\omega, \mathcal{C} \rangle \qquad (2.39)$$

The boundary operator and the exterior differentiation operator are 'adjoints' of each other.

We have already defined the homological invariant H_p as

$$H_p(M) = \frac{Z_p}{B_p}.$$

Here Z_p are the space of p-chains for which there are no boundaries and B_p are space of p-chains which are boundaries of some $p+1$ chain. We can write dual of this as $\frac{Z^p}{B^p}$ where Z^p are the space of closed forms ($d\omega = 0$) and B^p are the space of exact forms, $\omega = df$. The cohomology invariant is $H^p(M) = \frac{Z^p}{B^p}$

If $[\omega] \in \frac{Z^p}{B^p}$ and $[\mathcal{C}] \in \frac{Z_p}{B_p}$ then

$$\langle [\omega], [\mathcal{C}] \rangle = \int_\mathcal{C} \omega$$

If ω' and $\mathcal{C}'$ are in the same cohomology class and homology class, then

$$\langle \omega, \mathcal{C} \rangle = \langle \omega', \mathcal{C}' \rangle$$

Note: $\partial_p \partial_{p+1} = 0$, $d_{p+1} d_p = 0$.

Poincaré's Lemma If a space is homeomorphic to a point, then all closed forms are exact. This will make all cohomology groups trivial.

Examples

1. We realise $\mathbb{R}^n$ is contractible to a point. This is achieved by the map

$$\alpha : \mathbb{R}^n \otimes [0,1] \longrightarrow \mathbb{R}^n, \quad [x,t] \to (1-t)x$$

Using the above result and the Poincaré Lemma we have all closed forms on $\mathbb{R}^n$ are exact. Hence

$$H^p(\mathbb{R}^n, R) = 0, \quad n = 1, 2, \ldots n.$$

But interestingly we have $H^0(\mathbb{R}^n, R) = R$.
Proof Closed zero forms are functions f such that $df = 0$. But there are no forms lower than 0. The solution for $df = 0$ is $f = c$, constant. Hence for every number c, $-\infty < c < \infty$ there is an element of the zeroth cohomology class. Hence $H^0(\mathbb{R}^n, R) = R$

2. Consider $M = \mathbb{R}^2 - \{0\}$. Then an exact form is:

$$\omega = -\frac{y\,dx}{x^2 + y^2} + \frac{x\,dy}{x^2 + y^2}; \quad d\omega = 0.$$

If we look for an η such that $\omega = d\eta$, then we get,

$$\eta = \tan^{-1}\frac{y}{x} = \theta$$

But θ is defined as single valued only in $\mathbb{R}^2 - \mathbb{R}^+$. Hence ω is closed but not exact globally. Note ω is exact locally. That is, in any patch (chart) it is exact. But η and $\eta + C$ where C is a constant are exact. Hence $H^1(M, R) = R$.

3. Next we prove $H^1(\mathbf{S}^n, R) = 0$, $n > 1$. The same is true for all simply connected manifolds M. Since $\mathbf{S}^n$ is simply connected, every closed curve Γ can be contracted to the identity loop Γ^0 which is simply a point. We have for any closed one form

$$\int_\Gamma \omega = \int_{\mathbf{S}^1} \Gamma^*\omega = \int_{\mathbf{S}^1} \Gamma^0\omega,$$

where Γ^* is the induced map. Since we start with closed one form ω we have $\Gamma^*\omega = \Gamma^0\omega + d\eta$. But for us Γ^0 is a constant curve, it is 0:

$$\int_\Gamma w = \int_{\mathbf{S}^1} d\eta = 0.$$

2.6 Cohomology

Hence for all closed 1-forms on M, we have $\int_{S^1} w = 0$. Now we see that ω is also an exact form. This is easy, since we will have $\omega(x) = df(x)$ where

$$f(x) = \int_{x_0}^{x} w .$$

The above integral is independent of the path since the integral over any closed loop is zero. Hence $H^1(M, R) = 0$ for simply connected manifold M.

4. It is easy to show $H^1(\mathbf{S}^1, R) = R$. Remember $\mathbf{S}^1$ is multiply connected. Its fundamental group is the same as $\mathbb{R}^2 - \{0\}$.
5. Consider $\mathbb{R}^2$ after removing two points. Then, we can note $H^1(\mathbb{R}^2 - \{0, 1\}) = R \oplus R$. The number of holes will be counted.
6. For $M = \mathbf{S}^2$ we have already seen $H^1(\mathbf{S}^2, R) = 0$. We have to only compute $H^2(\mathbf{S}^2, R)$ since there are no forms beyond 2 for two-dimensional manifolds. To show this, we will first establish $H^2(\mathbf{S}^2, R) = H^1(\mathbf{S}^1, R)$. We have already seen $H^1(\mathbf{S}^1, R) = R$. First, we consider $S_+ = \mathbf{S}^2 - \{\text{North pole}\}$ (and similarly for $S_- = \mathbf{S}^2 - \{\text{South pole}\}$. Since both $S_\pm$ are contractible, any two form ω on them can be written as $d\eta_\pm$ with $\eta_\pm$ as one form. On the $S_- \cap S_+$ we have $\alpha = d(\eta_- - \eta_+) = 0$. Hence α is a closed one form on $S_- \cap S_+$. Consider two form $w = d\eta_\pm$, on $S_\pm$. This gives a one-to-one correspondence between closed two forms $S_- \cup S_+$ and closed one form on $S_- \cap S_+$. Hence, we also see one-to-one mapping between $H^2(\mathbf{S}^2, R)$ and $H^1(S_- \cap S_+, R)$. But $S_- \cap S_+$ is nothing but $\mathbf{S}^2$ with the north and the south pole removed (topologically a cylinder which is contractible to a circle). This space can be stereographically mapped to $\mathbb{R}^2 - \{0\}$. The mapping is given below (see Fig. 2.9):

$$z_i = \frac{2x_i}{1 + (x_1^2 + x_2^2)}, \quad i = 1, 2, \quad z_3 = \frac{1 - (x_1^2 + x_2^2)}{1 + (x_1^2 + x_2^2)}$$

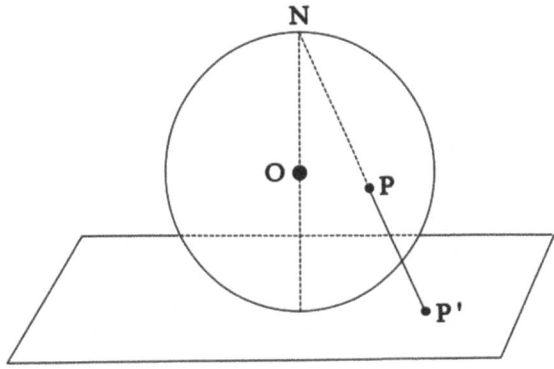

Fig. 2.9 Stereographic projection

We have $\sum z_i^2 = 1$ which defines a unit sphere $\mathbf{S}^2$. The north and south poles are identified with the origin and a point at infinity on $\mathbb{R}^2$. Hence we have

$$H^2(\mathbf{S}^2, R) = H^1(\mathbb{R}^2 - \{0\}, R) = H^1(\mathbf{S}^1, R) = R$$

7. The above result generalises to

$$H^m(\mathbf{S}^n, R) = 0, \; m > n \; \& \; 1 \leq m < n$$
$$= R, \; m = n, \; \& \; m = 0$$

Exercises

2.1 Consider ω a one form and $d\omega$ in 3 dimensions. Show this is related to 'curl' $\vec{\nabla} \times \vec{\omega}$ defined in vector calculus textbooks.

2.2 If ω is a two form in 3 dimensions then show that $d\omega$ is related to the divergence.

2.3 Write down Maxwell's equation in differential form notation.

2.4 Consider the addition of mass term to Maxwell's equations (known as Proca equation). Show that the equation describes spin-1 particle.

2.5 Show the equations for Kalb-Ramond fields describe massless scalar field.

2.6 Show Yang Mills field strength transforms covariantly under gauge transformations.

2.7 Show group manifold of $SU(2)$ is $\mathbf{S}^3$. Hence find

$$\pi_1(SU(2)), \pi_2(SU(2)), \pi_3(SU(2))$$

2.8 Find mapping between $\frac{SU(2)}{U(1)}$ and $\mathbf{S}^2$. Hence obtain the topological invariants of $\frac{SU(2)}{U(1)}$.

2.9 Consider the manifold $M = \mathbb{R}^3$ with a). A single Cartesian coordinate patch U_C which covers the whole space (x, y, z); b). A spherical coordinate patch U_S which has origin at $(x, y, z) = (0, 0, 0)$ and uses coordinates (r, θ, ϕ).

(i) Does the patch U_S cover the whole space? Why, or why not?
(ii) Write the coordinate transition functions between U_C and U_S. Are these functions differentiable?

Exercises

(iii) Write the coordinate basis vectors at a point $p \in M$ which has Cartesian coordinates (x, y, z). Expand these basis vectors in terms of U_S coordinate basis vectors, considering $p \in U_C \cap U_S$.

2.10 Consider a vector $\mathbf{V}$ at point p, given by $\mathbf{V} = V^\mu \frac{\partial}{\partial x^\mu}\big|_p$ and a function $f : M \to \mathbb{R}$ having the coordinate presentation $f(x, y, z) = x^2 + y^2 + z^2$. Evaluate $V(f)$. Will $V(f)$ remain the same when evaluated in U_S basis? Verify your claim by explicit calculation.

2.11 Consider the one form $\omega = xdx + ydy + zdz$ in M. Express ω using U_S one-form basis.

2.12 The current one-form is defined as $J = J_\mu dx^\mu = \rho dx^0 + j_i dx^i$, where ρ is the charge density and $\mathbf{j} \equiv (j_1, j_2, j_3)$ is the current density. Show that the other two Maxwell's equations: $\nabla \cdot \mathbf{E} = \rho$ and $\nabla \times \mathbf{B} = \mathbf{j} + \frac{\partial \mathbf{E}}{\partial t}$ follow from the equation $*d*F = J$, where '$*$' is the dual operator in Minkowski space.

2.13 Hamilton's equations in phase space, having coordinates (q_i, p_i) takes the form

$$\frac{dq_i}{dt} = \frac{\partial H}{\partial p_i} \qquad \frac{dp_i}{dt} = -\frac{\partial H}{\partial q_i}$$

where the Hamiltonian $H(q_i, p_i)$ is a smooth function in phase space. The set of curves $q_i(t), p_i(t)$ satisfying Hamilton's equations is called a **Hamiltonian flow**. An infinitesimal volume element in phase space is expressed $\omega = \sum dq_i \wedge dp_i$, also known as **symplectic form**. Show that Hamiltonian flows preserve the symplectic form.

2.14 Let $K = \{p_1, p_2, p_3, p_4, [p_1, p_2], [p_2, p_3], [p_3, p_4], [p_4, p_1]\}$ by a simplicial complex whose polyhedron is a square. Show that

$$H_0(K) \cong \mathbb{Z} \qquad H_1(K) \cong \mathbb{Z}$$

2.15 Consider the manifold $M = \mathbb{R}^3$ and a one-form $\omega = a_i dx^i$. Show that Stokes' theorem can be expressed as:

$$\int_S (\nabla \times \boldsymbol{\omega}) \cdot d\mathbf{S} = \oint_C \boldsymbol{\omega} \cdot d\mathbf{C}$$

where $\boldsymbol{\omega} = (a_1, a_2, a_3)$ (A vector in $\mathbb{R}^3$) and C is the boundary of a surface S. Similarly, for a two form $A = \frac{1}{2} A_{\mu\nu} dx^\mu dx^\nu$, show that Stokes theorem can be written as

$$\int_V \nabla \cdot \mathbf{A} \, dV = \oint_S \mathbf{A} \cdot d\mathbf{S}$$

where $A^\lambda = \epsilon^{\lambda\mu\nu} A_{\mu\nu}$ are the components of the vector $\mathbf{A}$ and S is the boundary of a volume V. (Note: $\epsilon^{\lambda\mu\nu} A_{\mu\nu}$ are the components of the three-dimensional Levi-Civita tensor).

2.16 Let $\mathbf{S}^1 = \{e^{i\theta} : 0 \leq \theta < 2\pi\}$. Compute the cohomology groups $H^0(\mathbf{S}^1)$ and $H^1(\mathbf{S}^1)$.

References

1. S. Lang, *Differential Manifolds* (Springer, Berlin, 1985)
2. H. Flanders, *Differential Forms with Applications to the Physical Sciences* (Dover Publications, Mineola, 1963)
3. M. Stone, P. Goldbart, *Mathematics for Physics* (Cambridge University Press, Cambridge, 2009)
4. M. Kalb, P. Ramond, Classical direct interstring action. Phys. Rev. **D9**, 2273 (1974)
5. C. Nash, S. Sen, *Topology and Geometry for Physicists* (Academic Press, New York, 1983); C. Nash, *Differential Geometry and Quantum Field Theory* (Academic Press, New York, 1991)
6. M. Nakahara, *Geometry, Topology and Physics* (Institute of Physics Publishing, London, 2003)

Riemannian and Pseudo Riemannian Geometry

As explained in the previous chapter we should distinguish terms like topological spaces, differential topology, differential geometry, algebraic topology and algebraic geometry. Since these objects appear in several places in quantum physics and also in the general theory of relativity, we make brief remarks about them.

1. Topological spaces are very generic structures involving open and closed sets with maps between them. They are characterised by several features like dimensions, compactness, boundaries, etc.
2. Differential topology introduces differentiability besides the continuity of the topological spaces. This also provides invariant features which brings in quantitative ideas. But still, it does not distinguish between a curvy line and a straight line or a circle from an ellipse.
3. Differential geometry brings out the features left out through the metric and distance between elements.
4. Algebraic topology provides the topological nature of spaces through abstract algebra. It was originally known as combinatorial topology. This focuses on constructing the topological space through simpler ones. The algebraic tool for such a construction is simplicial complex, which was used in the previous chapter to study invariants like homology and homotopy.
5. Algebraic geometry is the study of the spaces and the manifolds obtained through polynomial equations. They are the zeroes of the polynomials of several variables. The main focus is solutions of a collection of polynomial equations which are also known as algebraic variety. For example, we have been studying plane curves like straight lines, ellipses, parabolas, hyperbola or cubic equations etc in analytic geometry. We look for singular points, or those with special roles like intersections or crossings. These can be extended to complex equations and associated analyticity questions also.

Having seen the topological nature of differential geometry and smooth manifolds, we will proceed to a particular class of geometries known as Riemannian geometry. These and extensions of such geometries to pseudo-Riemannian ones play an important role in gauge and gravitational theories.

3.1 Riemannian Geometry

Here, we introduce an inner product on the tangent space which varies smoothly. The introduction of the inner product provides us with notions of angles between tangents, length of curves, area of surfaces and volumes. This geometry follows the axioms developed by Bernhard Riemann [1] given in his well-known lectures 'On the hypotheses which lie at the bases of geometry'. This generalised notions of the geometry of surfaces in $\mathbb{R}^3$ to higher dimensions as well as differentiable manifolds. This was further generalised to pseudo-Riemannian geometry, which enabled the formulation of Einstein's special and general theory of relativity.

3.1.1 Inner Product and the Metric

In $\mathbb{R}^3$ we can consider the inner product of two vectors

$$\vec{a} = (a^1, a^2, a^3), \quad \vec{b} = (b^1, b^2, b^3)$$

through

$$\vec{a} \cdot \vec{b} = a^1 b^1 + a^2 b^2 + a^3 b^3 . \tag{3.1}$$

This also provides the length of the vector as $|a| = \sqrt{\vec{a} \cdot \vec{a}}$. We can also write the inner product using metric[1] as

$$\vec{a} \cdot \vec{b} = g_{ij} a^i b^j , \tag{3.2}$$

where

$$g_{ij} = \begin{pmatrix} 1 & 0 & 0 \\ 0 & 1 & 0 \\ 0 & 0 & 1 \end{pmatrix} .$$

This leads us to use the metric in general curvilinear coordinates to provide the distance between two nearby position vectors $\vec{r}, \vec{r} + \vec{dr}$ as

$$ds^2 = g_{ij} \, dx^i \, dx^j . \tag{3.3}$$

---

[1] Einstein summation convention.

3.1 Riemannian Geometry

The above equation can be written in terms of the one form dx^i basis as

$$ds^2 = g_{ij} dx^i \otimes dx^j . \tag{3.4}$$

We can use the metric to relate contravariant and covariant vectors:

$$a_i = g_{ij} a^j .$$

Under a change of coordinates $x^i \to x'^i$, the metric transforms as:

$$g'_{ij} = \frac{\partial x^r}{\partial x'^i} \frac{\partial x^s}{\partial x'^j} g_{rs} . \tag{3.5}$$

These transformations imply, as expected, g_{ij} is a rank two covariant tensor. (*problem*: use cylindrical and spherical polar coordinates and extract the metric)

Definition 3.1 We also define contravariant second rank tensor g^{ij} as the inverse of g_{ij}. This gives

$$g^{ik} g_{kj} = \delta^i_j , \tag{3.6}$$

where δ^i_j is known as the Kronecker delta.

For scalar functions $f(x)$, we can define the gradient (or exterior derivative of scalar providing us one form):

$$df = \nabla f = \frac{\partial f}{\partial x_i} dx^i . \tag{3.7}$$

However, the differentiation of a vector field will not give us a second-rank tensor (except in Cartesian coordinates) and will require corrections provided by what are known as connections. This leads us to the concept of parallel transport.

3.1.2 Parallel Transport and Connection

Parallel transport achieves transporting a vector (or tensors) along a curve maintaining parallel property along the curve. This requires the introduction of a connection or a covariant derivative. This transports a vector at a point on a curve to another vector at the neighbouring point. There may be many connections or covariant derivatives along a curve but there is a special one known as Levi Civita connection coming from the metric. Consider a vector $V^i(x)$ with the transformation property

$$V^i(x) \to V'^i(x') = \frac{\partial x'^i}{\partial x^k} V^k(x) . \tag{3.8}$$

We want to make sure $D_j V^i(x)$ transforms like a tensor. It is a standard exercise in tensor calculus to see

$$D_j V^i(x) = \partial_j V^i(x) + \Gamma^i_{jk} V^k(x) \tag{3.9}$$

where Γ^i_{jk} is the Christoffel symbol added to cancel the additional terms that arise in $\partial_j V^i$ spoiling the tensorial property. For a given metric g_{ij}, the explicit form of the Christoffel symbol can be obtained imposing covariant derivative of the metric tensor vanishes:

$$\Gamma^i_{jk} = \frac{1}{2} g^{ir} \{\partial_k g_{jr} + \partial_j g_{kr} - \partial_r g_{jk}\} . \tag{3.10}$$

1. The connections can be more general than the above. For general relativity we focus on the above special connection.
2. Γ^i_{jk} does not transform as a third rank tensor as one would have expected.
3. It is useful to mention about the torsion tensor T^i_{jk} here. Torsion of a curve measures the twist along a curve about the tangent vector. It generalises the Frenet-Serret formula in $\mathbb{R}^3$. For a general connection (beyond Eq. (3.10) it is given by the antisymmetric part.

$$T^i_{jk} = \Gamma^i_{jk} - \Gamma^i_{kj} \tag{3.11}$$

With the torsion term even the covariant derivatives of the scalar field do not commute.

It is a convention to write the partial derivative $\partial_i V^j$ as V^j,i and covariant derivative $D_i V^j$ as $V^j;i$.

Having introduced the covariant derivative of a contravariant vector, we can obtain the covariant derivative of a covariant vector by contracting with constant contravariant vector. The result is

$$D_i V_j(x) = \partial_i V_j - \Gamma^k_{ij} V_k \quad \text{or} \quad V_{j;i} = V_{j,i} - \Gamma^k_{ij} V_k \tag{3.12}$$

From the expression for Christoffel symbol given (3.10) we can prove:

$$\Gamma^i_{jk} = \Gamma^i_{kj} \tag{3.13}$$

Using the fact that g^{ij} and g_{ij} are 'inversely' related we can show:

$$dg = g\, g^{ij}\, (dg_{ij}), \quad \partial_i g = g\, g^{jk}(\partial_i g_{jk}) , \tag{3.14}$$

where $g = Det g_{ij}$. The above relations are useful when varying Einstein-Hilbert action to obtain Einstein's equation.

3.2 Riemann Tensor

We will be interested in the curved spaces with curvature and other properties. They are required for both gauge theories and Einstein's theory of gravity. In $\mathbb{R}^n$ we have the partial derivatives which commute. That is,

$$\partial_i \partial_j f(x) = \partial_j \partial_i f(x), \quad [\partial_i, \partial_j] f = 0. \tag{3.15}$$

In curved space, we need the covariant derivatives to act on vectors and higher-rank tensors. For example, we need: $[D_i, D_j] V^k$. If this is zero, then we will see that the space is flat and has no curvature. Suppose the following action on a scalar function $f(x)$:

$$[D_i, D_j] f(x) = 0,$$

then the space has no torsion.

3.2 Riemann Tensor

Consider the commutator of two covariant derivatives acting on a vector field. It can be written as:

$$[D_i, D_j] V^k = R^k_{ijl} V^l \tag{3.16}$$

The new tensor R^k_{ijl} is known as the Riemann tensor and it measures the curvature of the space we are considering. It is a generalisation of curvature in two-dimensional surfaces.

When the connection is given by Christoffel symbols (3.10), we can easily obtain the expression for the Riemann tensor in terms of the metric and its derivatives. It is given by:

$$R^i_{jkl} = \partial_k \Gamma^i_{jl} - \partial_l \Gamma^i_{jk} + \Gamma^i_{js} \Gamma^s_{kl} - \Gamma^i_{ks} \Gamma^s_{jl}. \tag{3.17}$$

Symmetry Properties of Riemann Tensor $R_{ijkl} = g_{im} R^m_{jkl}$

1. $R_{ijkl} = -R_{ijlk}, \quad = -R_{jikl} = R_{jilk}$
2. $R_{ijkl} + R_{iklj} + R_{iljk} = 0$ (Bianchi identity I)
3. $R_{ijkl} = R_{klij}$
4. $R_{ijkl;m} + R_{ijlm;k} + R_{ijmk;l} = 0$ (Bianchi identity II)

Riemann tensor in N dimensional space has N^4 independent components before symmetries are imposed. But the symmetry properties reduce it to $\frac{N^2(N^2-1)}{12}$. In two dimensions, this reduces to 1. Hence, in the two-dimensional space we can write:

$$R_{ijkl} = C \left(g_{ik} g_{jl} - g_{il} g_{kj} \right), \tag{3.18}$$

where C is known as Gaussian curvature.

Given Riemann tensor R_{ijkl} we can define some useful tensors. Ricci tensor and scalar curvature are defined as:

$$R_{ij} = R^k_{ikj}, \quad R = R^i_i. \tag{3.19}$$

Another useful tensor is the Einstein tensor:

$$G_{ij} = R_{ij} - \frac{1}{2} g_{ij} R. \tag{3.20}$$

3.2.1 Geodesics

One of the most important questions that will come up is what is the least distance between two points in a Riemannian manifold. If we take a curve joining two points (denoted as 1, 2) the length of the curve between those points can be given as a line integral

$$s = \int_1^2 \sqrt{g_{ij} \frac{dx^i}{dt} \frac{dx^j}{dt}} \, dt, \tag{3.21}$$

where 't' parametrizes the distance along the curve from $1 \to 2$. To find the geodesic distance (the least distance) we have to minimise the above expression which becomes conventional variational problem in mechanics. To find that we vary the path and minimise the distance, keeping the end points fixed. This is similar to the action principle which leads to Euler Lagrange equation.

The Euler Lagrange equation for the action S:

$$S = \int dt \, g_{\mu\nu} \frac{dx^\mu}{dt} \frac{dx^\nu}{dt} = \int dt \, L \tag{3.22}$$

is

$$\frac{d}{dt}\left(\frac{\partial L}{\partial \dot{x}^\mu}\right) = \frac{\partial L}{\partial x^\mu} \tag{3.23}$$

We leave it as an exercise for the readers to show the geodesic curve satisfies the following equation:

$$\frac{d^2 x^\mu}{dt^2} + \Gamma^\mu_{\nu\rho} \frac{dx^\nu}{dt} \frac{dx^\rho}{dt} = 0 \tag{3.24}$$

3.3 Laplace Beltrami Operator

In the previous chapter, we introduced p forms and exterior differentiation through 'd'. The exterior differentiation on the scalar function is nothing but the gradient. We also come across in Euclidean geometry the Laplace operator ∇^2. We will now see how this generalises to curved manifolds in a coordinate independent way. Also, it plays a role as a self-adjoint operator in conventional quantum theory. We will explore these questions since they have an important bearing in quantum physics as operators on a Hilbert space of square-integrable functions.

Recall, exterior d acting on a p form gives $p+1$ form. We also defined a boundary operator δ acting on a p give a $p-1$ form. This is defined through the Hodge star operator which maps p form to $n-p$ form. Remember both p and $n-p$ forms have the same number of components through the identity $^n C_p = {}^n C_{n-p}$

Remember the Hodge star operator defined in Chapter 2. For a p-form ω, we define Hodge star operation as follows:

$$*\omega = \frac{\sqrt{g}}{p!(n-p)!}\omega_{\mu_1\mu_2..\mu_p}\epsilon^{\mu_1\mu_2..\mu_p}_{\nu_{p+1}\nu_{p+2}..\nu_n}dx^{\nu_{p+1}} \wedge dx^{\nu_{p+2}} \ldots \wedge dx^{\nu_n}. \quad (3.25)$$

This mapping can be obtained through the metric and inner product of two arbitrary p-forms α, β

$$<\alpha,\beta> = \int \alpha \wedge *\beta. \quad (3.26)$$

This star operator uses the Levi Civita symbol $\epsilon^{\mu_1\mu_2\cdots}$ and the metric $g_{\mu_1\mu_2}$.

The above is described by constructing a Hilbert space from p forms through the inner product.

$$<\alpha,\beta> = \int_M \alpha \wedge {}^*\beta = \frac{1}{p!}\int_M \sqrt{g}d^n x \alpha_{i_1i_2..i_p}\beta^{i_1i_2...i_p}. \quad (3.27)$$

Such an inner product can be used to define 'adjoint' $\delta = d^\dagger$. This is defined through

$$<d\alpha,\beta> = <a,\delta\beta>. \quad (3.28)$$

Note that the operator δ takes a p form to $p-1$ form.

Now, we define a second-order differential operator generalising the Laplace operator in Euclidean space as:

$$\Delta = (d\delta + \delta d). \quad (3.29)$$

This maps a p form to p form and has the well-known form in Euclidean space $-\nabla^2$. This is called the Laplace Beltrami operator. It is formally a self-adjoint positive operator on $\mathcal{L}^2$ functions on the manifold M (which is a compact manifold without boundary). On manifolds with boundaries, one needs boundary conditions and it can spoil the self-adjointness. These are important questions whose answers were developed by von Neumann [2] which we explain now. But in our study of classical electrodynamics, we are familiar with Dirichlet (functions vanish on the boundary) or Neumann (normal derivatives vanish on the boundary). The general analysis is due to von Neumann.

3.4 Self Adjoint Operators

Consider Hilbert space $\mathcal{H}$ of square-integrable functions $\mathcal{L}^2$. A bounded operator $\mathcal{O}$ is self-adjoint if

$$<x|\mathcal{O}\,y> \;=\; <\mathcal{O}\,x|y>, \forall x, y \in \mathcal{H}. \tag{3.30}$$

For an unbounded operator, the above equation is not enough. Most of the operators in quantum theory are unbounded ones.

An unbounded operator $\mathcal{O}$ is a symmetric or hermitian operator if

$$<x|\mathcal{O}\,y> \;=\; <\mathcal{O}\,x|y>, \; for\; all\; x, y \in \mathcal{D}(\mathcal{O}), \tag{3.31}$$

where $\mathcal{D}(\mathcal{O})$ denotes the domain of the operator $\mathcal{O}$. In order to prove the hermiticity of the operator, we do not need the adjoint $\mathcal{O}^\dagger$ of the operator. We need to only verify (3.31) is in the domain $\mathcal{D}(\mathcal{O})$.

An unbounded operator $\mathcal{O}$ with domain $\mathcal{D}(\mathcal{O})$ can be self-adjoint only if it satisfies:

1. $<x|\mathcal{O}\,y> \;=\; <\mathcal{O}\,x|y>\;$ for all $x, y \in \mathcal{D}(\mathcal{O})$
2. $\mathcal{D}(\mathcal{O}) = \mathcal{D}(\mathcal{O}^\dagger)$

The first condition defines a symmetric operator. We are interested in self-adjoint operators since they represent observables in quantum theory and will have real eigenvalues and any state in the Hilbert space can be expressed as linear combination of eigenvectors of such an operator.

Remark In physics textbooks, it is taken for granted hermitian and self-adjointness as the same. But there are differences which are due to the unboundedness of the operator and lead to interesting consequences. Self-adjointness implies hermiticity but not vice-versa.

We might be able to change the boundary conditions and extend the domain of the hermitian operator $\mathcal{D}(\mathcal{O})$ so that the operator becomes self-adjoint in the new domain. In that case, the operator is known as essentially self-adjoint.

3.4 Self Adjoint Operators

Given the above definitions, von Neumann obtained conditions to check the self adjointness [3].

Consider $\mathcal{O}^\dagger$ and solve the eigenvalue equation

$$\mathcal{O}^\dagger u = \pm iu \qquad (3.32)$$

Let $n_\pm$ be the number of independent eigenvectors with eigenvalues $\pm i$. They are called the deficiency indices.

Theorem 3.1 *The operator is self-adjoint or essentially self-adjoint or not self-adjoint if the deficiency indices satisfy certain conditions. If*

1. $n_+ = n_- = 0$, *then the operator is self-adjoint.*
2. $n_+ = n_- = n$, *then the domain of the operator $\mathcal{O}$ can be extended so that it is essentially self-adjoint.*
3. $n_+ \neq n_-$, *then the operator is not self-adjoint and cannot be extended either.*

The above difficulties arise when we consider manifolds with boundaries. We will explain the theorem with simple examples.

Examples

1. Consider a conventional free particle in a line interval $[0, 1]$. This is nothing but the usual particle in a box given in the introductory quantum mechanics textbooks. The Hamiltonian is $H = -\frac{d^2}{dx^2}$ and the momentum is $\hat{p} = -i\frac{d}{dx}$. The momentum operator $\hat{p}$ is a symmetric operator. But the number of independent solutions to the equation

$$\hat{p}\,\psi(x) = \pm i\psi$$

   indicates the deficiency index to be $(1, 1)$ (Exercise 7). Following von Neumann's conditions, we find the momentum operator can admit self-adjoint extension if $\psi(1) = e^{i\theta}\psi(0)$. The parameter θ gives a family of one parameter extensions for the operator.

2. For the Hamiltonian $H = -\frac{d^2}{dx^2}$ the deficiency indices will turn out to be $(2, 2)$. There are 4 parameter extensions of the operator H following von

(continued)

Neumann's theorem to ensure it to be essentially self-adjoint. These will be reflected as possible boundary conditions.
3. Consider free particle along positive x-axis $x > 0$. Show the momentum operator $\hat{p} = -i\frac{d}{dx}$ has indices (0, 1), is not self-adjoint and cannot be extended to make it self-adjoint since $n_+ \neq n_-$. But the free Hamiltonian $H = -\frac{d^2}{dx^2}$ has indices (1,1). We can give a suitable boundary condition at $x = 0$ and make it self-adjoint.

The previous analysis for one-dimensional manifolds can be extended to higher-dimensional ones using the Laplace Beltrami operator $-\nabla^2$.

3.4.1 Self-Adjoint Laplace Beltrami Operator on a Manifold with Boundary

Von Neumann's theory of deficiency indices for self-adjointness is not always useful when the indices turn out to be infinite. This happens in the following cases:

Consider an arbitrary Riemannian manifold $\mathcal{M}$ with boundary $\partial\mathcal{M}$. We are interested in solving the eigenvalue equation

$$-\nabla^2 \Phi = \lambda \Phi \tag{3.33}$$

with boundary values $\Phi_{\partial\mathcal{M}}$ and $\partial_n \Phi_{\partial\mathcal{M}}$ where ∂_n stands for normal derivative of the scalar function Φ at the boundary. Laplace operator has infinite number of extensions to make it essentially self-adjoint given by the following conditions:

$$(\Phi + i\,\partial_n\Phi))_{\partial\mathcal{M}} = U\,(\Phi - i\,\partial_n\Phi))_{\partial\mathcal{M}} \tag{3.34}$$

where the unitary matrix U links the two functions. The above conditions were provided by Alberto et al. [4].

It is easy to see the conventional Dirichlet and Neumann conditions are two extreme limits of the above conditions. In Chap. 5, we will use the above result in the specific manifold $\mathbb{R}^2$ − Disc. Robin boundary condition is defined as interpolating the Dirichlet and Neumann conditions. It leads to interesting bound states localised near the boundary with applications to several areas of quantum physics like Casimir energy, quantum Hall effect, blackhole bound states, etc. [5].

We will see the implications of the above analysis in the Chaps. 4–6.

3.5 Pseudo Riemannian Geometry

We studied in the previous section, Riemannian manifolds which are characterised by positive definite metric structure. This ensures the distances are positive. But, special and general relativity needs a generalisation of not positive definite metric. This is signalled by positive and negative eigenvalues of the metric. The difference between them is called the signature. For example the Minkowski space has metric[2] $g_{\mu\nu} = Diag(1, -1, -1, -1)$ with the signature 2. The determinant is negative too. General relativity too uses such a metric. Minkowski space can be termed as a pseudo Euclidean space. General pseudo-Riemannian manifold is characterised locally at every spacetime event a tangent space which is pseudo-Euclidean viz., Minkowski spacetime.

The geometric structures like covariant and contravariant vectors, tensors etc., introduced for Riemannian manifold generalises automatically to these also. The new input that will be required comes from the square root of determinant $\sqrt{g}$ of the metric. Since the metric determinant is negative we have to take $\sqrt{-g}$. The rest of the steps are similar to Riemannian manifolds.

With the above definition of pseudo-Riemannian manifolds, we can ask how we generalise the Newtonian gravitational equations for general relativity for some matter distribution. Recollect in the Newtonian theory of gravity we have for matter described by a density $\rho(x, y, z)$ and the gravitational potential $V(x, y, z)$ is given as solutions of Poisson's equation:

$$\nabla^2 V = 4\pi \mathcal{G} \rho \tag{3.35}$$

where $\mathcal{G}$ denotes the Newton's gravitational constant.

In Einstein's theory, the matter distribution is replaced by energy-momentum tensor $T^{\mu\nu}$ which curves the spacetime given by pseudo-Riemannian metric $g_{\mu\nu}$. Also, the equations should be generally covariant and naturally generalises to

$$G_{\mu\nu} = \left(R_{\mu\nu} - \frac{1}{2} g_{\mu\nu} R \right) = 8\pi \mathcal{G} \, T_{\mu\nu} \tag{3.36}$$

where we have assumed the unit that the velocity of light $c = 1$ For weak Gravitational field we assume

$$g_{\mu\nu} = \eta_{\mu\nu} + h_{\mu\nu} \tag{3.37}$$

where $h_{\mu\nu}$ are taken to be small (i.e $h_{\mu\nu}^2 \approx 0$). The above equations lead to the nonrelativistic limit, to Poisson's equation. The metric in that limit will be

$$ds^2 = g_{\mu\nu} dx^\mu dx^\nu = (1 - 2\Phi) dt^2 - (1 + 2\Phi) dr^2 - r^2 d\Omega^2 \tag{3.38}$$

where Φ is the scalar potential. With factors c included, it simplifies even further.

---

[2] We use the units that the velocity of light $c = 1$.

3.5.1 Einstein Hilbert Action

The next question is: Can the gravitational field equation given in Eq. (3.36) be obtained from an action principle? For that, we need to integrate over the space-time manifold by using a density. In Minkowski space we integrate the scalar function. But in a curved manifold we need scalar density obtained by including $\sqrt{-g}$ to perform the integration. The volume element is $\sqrt{-g}d^4x$. Without this additional term, the action will not be invariant under general coordinate transformations. This is because the volume measure d^4x is not invariant but it is $\sqrt{-g}d^4x$ serves as a proper invariant measure. In Minkowski space, the addition will not affect since $g = -1$.

Given the above principle, the action known as Einstein Hilbert action [6] can be written down which leads to Einstein's field equations. It is

$$S_{Gravity} = \frac{1}{16\pi G} \int \sqrt{-g}\, R d^4x \qquad (3.39)$$

To this, we have to supplement the matter part. The Euler-Lagrange equation for the above action leads to Einsteins Field equation. (Hint: Use the variation of the Determinant of the metric g.)

3.5.2 Rindler Spacetime

The Minkowski space is the flat space representing inertial frames in special relativity. The simplest extension which produces curved spacetime is the uniformly accelerated frame. In nonrelativistic physics, we consider for example a uniform gravitational field close to the surface of the earth. The corresponding situation in relativistic physics leads to the uniformly accelerated non-inertial frame which can be compared with the constant gravitational field. This produces intriguing and at the same time carrying features like event horizon which characterises black hole solutions. We will briefly describe the new metric.

Let us start with the Minkowski metric given by:

$$ds^2 = dT^2 - dX^2 - dY^2 - dZ^2 \qquad (3.40)$$

Consider transformations given by uniformly accelerated motion defined by:

$$X = x\cosh t,\ T = x\sinh t,\ Y = y,\ Z = z \qquad (3.41)$$

This gives $\tanh t = \frac{T}{X}$, $x^2 = X^2 - T^2$. Substituting the transformation in the Minkowski metric we get the metric for uniformly accelerated metric as:

$$ds^2 = (x^2)dt^2 - dx^2 - dy^2 - dz^2 \qquad (3.42)$$

Fig. 3.1 Rindler spacetime

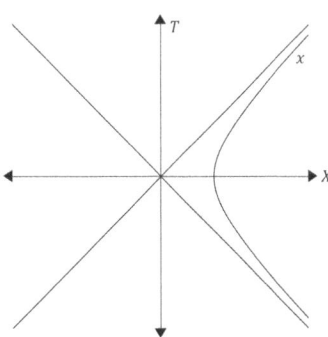

This defines the Rindler metric [7] where the observer is at rest in the non-inertial frame experiencing the uniform gravitational field. Rindler observer is at constant X. It is not a geodesic as can be checked by the equations. It is as mentioned earlier, a uniformly accelerated one. Rindler observers cannot see events from $X < 0$. Since metric is independent of t any translation in t is a symmetry and the operator $\frac{\partial}{\partial t}$ is known as Killing vector. If we transform back to X, T we can show it is a boost.

Rindler coordinates (Fig. 3.1) have a 'coordinate singularity' at $x = 0$. The determinant of the metric becomes zero at this point. We can consider this as the 'Rindler horizon'. None of the Rindler observers who have a constant position will ever see light signals from $T > X$. This feature resembles that of a horizon of Schwardschild's blackhole solution.

Exercises

3.1 Obtain the expression for Christoffel Γ^i_{jk} in terms of the metric g_{ij}.

3.2 Find the transformation rule for the connection coefficients Γ^i_{jk} between two different coordinates explicitly, thereby showing that they do not transform like a tensor.

3.3 For a two-dimensional manifold, the Riemann curvature tensor can be written in the form:

$$R_{ijkl} = C(g_{ik}g_{jl} - g_{il}g_{kj})$$

where C is a smooth function on the manifold, known as Gaussian curvature. Evaluate the Ricci tensor and Ricci scalar.

3.4 Show

$$dg = g\, g^{ij}\, (dg_{ij}), \quad \partial_i g = g\, g^{jk}(\partial_i g_{jk}) \tag{3.43}$$

where $g = Det g_{ij}$.

3.5 Show: $\partial_i \log \sqrt{g} = \frac{1}{2g}\partial_i g$ and $\Gamma^i_{ji} = \frac{\partial \log \sqrt{g}}{\partial x_j}$, $D_i \sqrt{g} = 0$.

3.6 Show that the symmetry of the Christoffel symbol is related to zero torsion condition.

3.7 Show that the geodesic curve satisfies the equation:

$$\frac{d^2 x^\mu}{dt^2} = \Gamma^\mu_{\nu\rho} \frac{dx^\nu}{dt} \frac{dx^\rho}{dt} \tag{3.44}$$

3.8 Prove the straight line provides the least distance between two points in $\mathbb{R}^3$. Obtain the curve for the extremal distance on the surface of a sphere.

3.9 Consider the Poincare upper half plane $U = (x, y) \in \mathbb{R}^2$, $y > 0$. This space is endowed with the metric tensor

$$ds^2 = \frac{dx \otimes dx + dy \otimes dy}{y^2} \tag{3.45}$$

Find the geodesic equations and solve them to show that the geodesics are either semi-circles or vertical lines in the $x - y$ plane with $y > 0$.

3.10 Show that the number of independent solutions to

$$\hat{p}\, \psi(x) = \pm i\psi$$

on an interval $[0, 1]$ is $(1, 1)$.

3.11 For the Hamiltonian $H = -\frac{d^2}{dx^2}$, on an interval show that the deficiency indices are $(2, 2)$

3.12 Show Einstein's equations lead in the non-relativistic limit (when velocities are small and the gravitational field is weak) to the Poisson's equation. Obtain the approximate metric in that limit.

3.13 Show that the Euler Lagrange equation for the Einstein-Hilbert action gives Einsteins Field equation. (Note we have already provided the variation of the Determinant of the metric g.)

References

1. B. Riemann, On the hypothesis which lie at the bases of geometry. Translated by William Clifford. Nature **VIII**(183), 14 (1998)
2. J. von Neumann, *Functional Operators. The Geometry of Orthogonal Spaces*, vol. 2 (Princeton University Press, Princeton, 1950)
3. D.M. Gitman, I.V. Tutin, B.L. Voronov, *Self Adjoint Extensions in Quantum Mechanics* (Birkhauser, Basel, 2010)
4. M. Asorey, A. Ibort, G. Marmo, Global theory of quantum boundary conditions and topology change. Int. J. Modern Phys. A **20**(5), 1001–1025 (2005)
5. T.R. Govindarajan, R. Tibrewala, Novel black hole bound states and entropy. Phys. Rev. D **83**, 124045 (2011)
6. D. Hilbert, Die Grundlagen der Physik. Found. Phys. (in German) **3**, 395 (1915); R.M. Wald, *General Relativity* (University of Chicago Press, Chicago, 1984)
7. W. Rindler, Hyperbolic motion in curved space time. Phys. Rev. **119**, 2082 (1960)

Part II

Topological Understanding of Defects in Crystalline Structure

4

We have studied in the first three chapters important lessons of topology and geometry. These are required to understand applications in relativistic and nonrelativistic Physics. In the forthcoming chapters till Chap. 6, we will study some applications in classical and quantum physics. Symmetries play an important role in phase structures of solids and liquids. There are also transitions between them which arise due to change of symmetries from one form to another. The crystalline structure of solids made up of atoms and molecules follow discrete group symmetries. There are also defects in these structural arrangements. Homotopy and homology provide a natural setting for understanding and classifying these defects in crystalline structures or ordered media. These descriptions play an important role in providing insights into the actual laboratory outcomes. The density of the defects affects the properties of materials in several ways including the elastic properties of materials. Hence, this chapter will focus on understanding the topological aspects due to defects in the crystalline structure.

4.1 Point Defects, Line Defects, and More

In a crystal, we have periodic arrangements of atoms. They are described by symmetry groups. A defect in a crystal is reflected in the microscopic region through the local arrangements being different from the rest of the crystal. There could be point defects, line defects or surface defects [1]

- Point defects can be due to vacancies (Fig. 4.1). It can also be an interstitial defect where an extra atom squeezed in the lattice configuration, or Schottky and Frenkel defects where ions are missing or included in a local region. The ions play a crucial role in the electrical conductivity of the material as well as optical properties. They also play a role in the thermodynamic equilibrium of

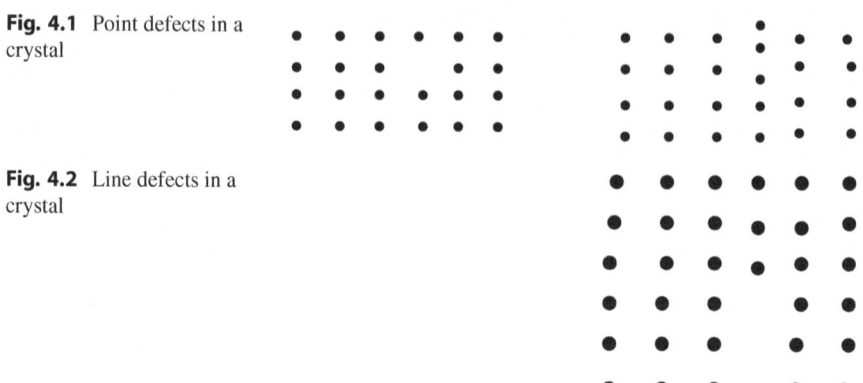

Fig. 4.1 Point defects in a crystal

Fig. 4.2 Line defects in a crystal

the substance. Hence, the density of such defects will affect various properties of solids.
- Dislocations are misalignments of crystalline structure which is a stringlike continuous defect. They are also known as line defects and are always present in real systems. They are needed to understand the strength and growth of crystals as illustrated in Fig. 4.2.

Density of defects can be controlled by temperature and pressure. The properties of materials can be modified through defects.

With this brief discussion on the role of defects in real materials, we will now consider ordered medium which assigns a function $\phi(x, y, z)$ to every point in the region. In other words, it maps from $\mathbb{R}^3$ to $\mathbb{R}$ or $\mathcal{C}$ (or to a topological space V). These maps are called order parameter spaces. If the map is simply a constant then we will consider the medium to be uniform. There could be other maps which are varying through the region. Such maps may end up in singularities at isolated points, curves or surfaces. These singular regions of dimensions zero, one and two are referred to as point, line and surface defects respectively.

We will study the order parameter space and possible singularities in them. The order parameter spaces are different from the examples given initially through crystals. These changes arise due to the breaking of the translational symmetry which affects the properties of the materials. Here, our attempt is to study the topological defects described by the singularities of the continuous function. We will relate such singularities to the topological invariants described earlier [2].

Now we will provide some examples of the order parameter space and explain how they arise in solid state systems [3].

4.1 Point Defects, Line Defects, and More

Examples

1. **Ferromagnetic liquid:** Ferromagnetism is described by magnetic moments of the spins of the atoms in a lattice interacting with each other. For simplicity, we take nearest neighbour interaction. That is, the Hamiltonian is

$$H = J \sum \vec{s}_i \cdot \vec{s}_{i+1} \tag{4.1}$$

where $\vec{s}_i$ is the spin at site i. In the continuum limit we have spins $\vec{s}$ in $\mathbb{R}^3$ pointing in different directions. But the spins at each point obey the relation $(\vec{s})^2 = 1$. Such a model will be applicable if the strength of local magnetisation is more than the lattice distances. This happens in the field theory approximation of the Heisenberg model of Ferromagnetism where nearest neighbor interaction between spin magnetic moments is considered. Then in that case, the function is a unit vector which varies over the three-dimensional space.

$$\vec{s}(x, y, z) = s_x(x, y, z)\hat{i} + s_y(x, y, z)\hat{j} + s_z(x, y, z)\hat{k} \tag{4.2}$$

with the condition $s_x^2 + s_y^2 + s_z^2 = 1$. The order parameter space is, therefore, the unit sphere $\mathbf{S}^2$.

2. **2D:** In planar two dimensions ($\mathbb{R}^2$) distribution of spins can be characterised by the function

$$\vec{s}(x, y) = \cos\phi(x, y)\hat{i} + \sin\phi(x, y)\hat{j} \tag{4.3}$$

A well-known example is superfluid He^4, which is described by a complex field

$$\Psi(x, y) = \Psi_0 \, e^{i\phi(x,y)} \tag{4.4}$$

3. **RP²:** An interesting modification of the above spin systems is the Nematic liquid crystals. Nematic liquid crystals are translucent liquid that changes the polarization of the electromagnetic wave when it passes through it. The molecules align in threadlike shape which explains the origin of the term (Nemato means thread in Greek). These are used in LED displays. Mathematically, the distribution of the molecules corresponds to the spins without an arrowhead. Hence the order parameter space is the set of unit vectors $\vec{n}(x, y, z)$ without its arrowhead. This makes the order parameter space to be given by the unit vector $\vec{n}$ and its 180° rotation identified. This is also known as projective plane **RP²**. It can also be understood as a sphere $\mathbf{S}^2$ in $\mathbb{R}^3$ with diametrically opposite points identified as one and the same. This is known as the projective plane $\mathbf{RP}^2 = \frac{\mathbf{S}^2}{\mathbb{Z}_2}$. Since it is

made of unit vector $\vec{n}$ without its arrowhead, we can characterise them using a traceless matrix M_{ij} constructed out of these n_i's as follows:

$$M_{ij} = \left(n_i\, n_j - \frac{1}{3}\delta_{ij} \right).$$

4. **Superfluid He^3**: This is an interesting example where the order parameter space is a more complicated generalization of **RP^2**. Superfluid He^3 exists in different phases out of which the one known as dipole locked A phase is what we explain. It is given by **RP^3**. This needs a three sphere S^3 as subspace in 4 dimensions given by $\sum_{i=1}^{4} X_i^2 = 1$. The order parameter can be described by a pair of orthonormal vectors

$$\vec{n}^1 \cdot \vec{n}^2 = 0, \qquad \vec{n}^i \cdot \vec{n}^i = 1, \qquad i = 1, 2 \tag{4.5}$$

We can also describe the same in complex notation by defining:

$$\vec{z} = \vec{n}^1 + i\, \vec{n}^2 \tag{4.6}$$

With this the space is given by

$$\vec{z} \cdot \vec{z}^* = 2, \quad \vec{z}.\vec{z} = 0 \tag{4.7}$$

Since there is no symmetry, which leaves a pair of orthonormal vectors unchanged the order parameter space is the full rotation group involving orthogonal matrices known in the literature as $SO(3)$ group.

4.1.1 Defects and Topology

Now we will see how the defects appear due to the topology of the order parameter space. Consider the example of spins in a two-dimensional plane. The field $\vec{s}(r, \theta)$, a unit vector, defines the order parameter. Suppose $\vec{s}(r, \theta)$ is continuous everywhere except at some point P. As we move $\vec{s}(r, \theta)$ on any circle around the singular point P, we require the field to obey

$$\vec{s}(r, \theta + 2n\pi) = \vec{s}(r, \theta),$$

where n denotes the winding number. We have illustrated in Fig. 4.3, winding numbers 1, 2, 3 are associated with different functional forms of the order parameter.

Note that, as we shrink the circle around the point, the winding number will remain unchanged even though the functions can change continuously. Thus, the winding number is invariant and interestingly captures the topology of the space even if the enclosing circle is far away from P. Intuitively, such an observation can

4.1 Point Defects, Line Defects, and More

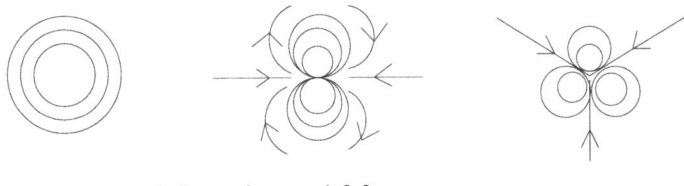

winding numbers   1, 2, 3

Fig. 4.3 1,2,3-Winding number

be stated as : '*We cannot comb a hairy ball. There will be at least one hair standing out.*'

If the singularity needs to be avoided the winding number should be zero. It is a necessary condition to escape singularity at P. But it is not sufficient. Since close to the point the vector field can be pinched and create a fork. This will result in a discontinuity in slope. But this discontinuity can be easily smoothened leaving the point without singularity. This smoothening can be done locally without affecting the fields far away.

Now, we provide the topological interpretation of the same. The order parameter space is a circle. This is because, for spin, we provide a constant magnitude vector along with an angle to indicate the arrow. When we provide a vector for the order parameter along a curve in $\mathbb{R}^2$, we are mapping the loop in $\mathbb{R}^2$ to a circle $\mathbf{S}^1$. The winding number n is simply the number of times the angle wraps around the loop. For visual effects, this can be imagined as a rubber band wrapped around a cylinder. Two maps with differing winding numbers cannot be deformed to one another by continuous local transformations. We will be forced to cut the rubber band and join again. This continuous deformation of the maps is the topological question as discussed in the first chapter.

As defined earlier two mappings of the order parameter space into the $\mathbf{R}^2$ will be homotopic if they can be deformed into each other. They are classified by the winding number of the maps. Winding number zero maps are the only ones which can be easily contracted to the point. This reflects there will not be singularity at P for such maps. For all other maps, there will be a singularity reflecting the origin of the defect in such descriptions. If there is a fork at P for zero winding number maps it can be removed by continuous deformations. Hence, they are topologically unstable singularities. But for nonzero winding numbers, the singularity cannot be removed and is known as topologically stable. Also, maps with the same winding number singularities can be deformed into each other. They are known as topologically equivalent.

Further, if there are two singular points X, Y with winding numbers n and m, we can construct a new deformed map which encircles each so that the new map shows winding number $n + m$ as explained in the Fig. 4.4. This makes a configuration of spins with winding number n and $-n$ (See Fig. 4.5) will be equivalent to trivial map. We get the new method to remove a singularity, which we can bring from far away an anti winding singularity. All the above characteristics establish the additive nature

Fig. 4.4 n+m winding number

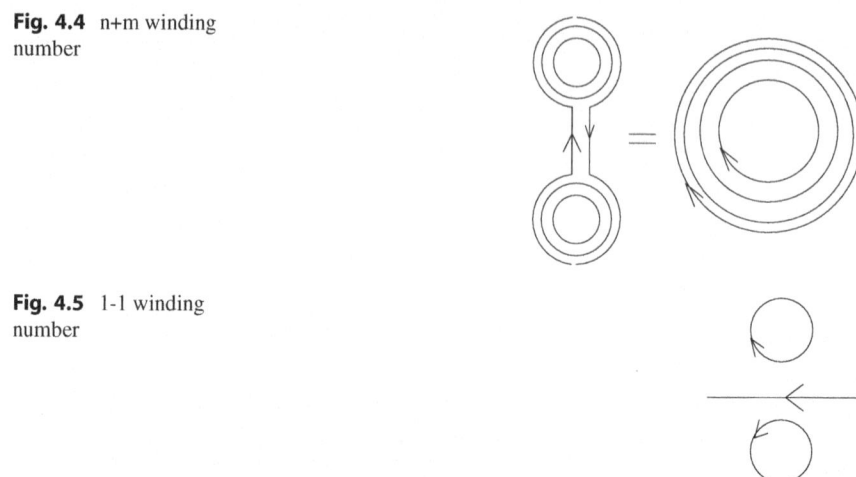

Fig. 4.5 1-1 winding number

of these defects from which we can easily see its connection to the fundamental group of the order parameter space. Note fundamental group can in principle be non-abelian and for the current example it is abelian in nature.

We can extend the above arguments by considering ordinary spins in 3 dimensions. The order parameter space is a sphere. These are mapped into the plane. If there are defects in this mapping then it has to be necessarily unstable. This is easy to see since any loop on a sphere can be continuously shrunk to a point. (A rubber band around an orange can be easily removed.)

It is clear from the above the homotopic properties of maps from the loops or spheres in the real space to the order parameter space classify the defects. These are given by fundamental group or higher homotopic groups. We discussed several examples of spaces and their fundamental and higher homotopic groups earlier. The symmetry group of transformations on the order parameter plays an important role in this. We now consider the action of this group of transformations on the order parameter space.

4.1.2 Group Theory and Order Parameter Space

Brief Remarks About Continuous Groups

A continuous group is one with elements labelled by several continuous parameters. It has group multiplication properties as well as structures to provide the notions of continuity to the open and closed sets of group elements. It also may be endowed with the metric structure with geometric properties. Since we are concerned with groups of transformations it will be obvious to note the meaning of two transformations that are close to each other. The two structures namely the group and topological properties are intertwined. We will usually be interested in Lie groups, which have stronger conditions about the structures near identity which are labelled

4.1 Point Defects, Line Defects, and More

by few parameters. All the condensed matter, statistical physics as well as quantum field theory examples of interest we study are provided by the Lie Groups.

1. Given a group G and $\tilde{G}$ a subset of elements connected to the identity element then $\tilde{G}$ is a normal subgroup.
2. The disconnected elements of G are cosets of the group. They are connected by some discrete group.
3. We have a group structure on the disconnected sets of group elements and these are the cosets of the normal subgroup $\tilde{G}$. They can be given a group structure also known as quotient group $\frac{G}{\tilde{G}}$. The product of two cosets is the coset, which contains the element obtained by multiplying any two elements of the two cosets.
4. The quotient group $\frac{G}{\tilde{G}}$ can also be seen to be the $\pi_0(G)$, the zeroeth homotopy group of G.

Groups and Order Parameter Spaces

Order parameter spaces always have group of transformations which link any two different points. This map from one point to another on a space by the group element need not be unique. There may be many elements which perform this mapping. But we are interested in the subgroup $\tilde{G}$ which leaves a point 'x' undisturbed. This is also known as a little group of x. It is also known as the isotropy subgroup of x. For example, if we have spin vectors $\vec{s}$, then the rotations in the plane normal to this axis will leave this vector fixed. This subgroup is not usually a normal subgroup. For example, the subgroups of rotations which leave different spin vectors fixed are not the same. For this reason, we choose a fiducial order parameter and obtain the little group of that point. For example, we choose spin pointing along the z-axis. This arbitrary choice will not affect the topological structure we are looking at. The quotient coset we obtain by considering space of cosets $\frac{G}{\tilde{G}}$ will be our order parameter space of study. Though the description looks mathematical it captures the structures arising from broken symmetry.

Consider for example a uniformly ordered medium and let the fiducial point x be the value everywhere due to the energetics. If the symmetry is the complete G then this point can be transformed to any other value. Again due to energetics a subgroup $\tilde{G}$ may be convenient. This will be reflected as a symmetry broken phase and $\tilde{G}$ will be the symmetry of the ordered phase.

Examples

We now consider several examples of order parameter spaces given earlier as cosets of Lie groups.

Examples

1. Spins in a plane. Here configurations are space of unit vectors in a plane. The group is SO(2). Only a trivial identity element will leave a vector invariant. Hence, the order parameter space is SO(2) itself. If we had taken O(2) as the configurations including reflections (2×2 orthogonal matrices with determinant ± 1), then subgroup H would be $\mathcal{Z}_2$. This would make the order parameter space a circle, which is the group manifold $SO(2)$ itself. If configurations are given as translations T then the subgroup will be translations by 2π will get identified leading again to the same order parameter space.
2. Spins in 3D. Here the configurations are given by $SO(3)$ consisting of rotations of a vector. But the subgroup that leaves the fiducial vector along z-axis invariant is $SO(2)$. Hence the quotient space $\frac{SO(3)}{SO(2)}$ is the order parameter space. This is nothing but S^2. But SU(2) group is the universal cover of $SO(3)$ (and acts transitively on the configurations) and the subgroup which leaves z-axis fixed is given by $H = \exp i\sigma_z \phi = U(1)$. This leads to the same $\mathbf{S}^2$.
3. Nematics and Biaxial nematics. Here, we have the molecule along a line and the configurations are again all possible rotations $SO(3)$. For nematic crystals, we have rotations molecule as well as 180° rotation of the molecule (perpendicular axis). This subgroup is known as D_h. For Biaxial nematics the subgroup turns out to be D_2 which is also known as the space of quaternions Q.
4. There are several discrete subgroups of $SU(2)$ which reflect the symmetries of molecules and the ordered phase will bring out the corresponding topological properties of these parameter spaces.

Homotopy Groups of Lie Groups and Coset Spaces

Since Lie groups have differential geometric and topological structures, we can obtain the homotopy groups of such groups as well as the coset spaces obtained by quotienting them with subgroups. For example, $U(1)$ group is nothing but the circle S^1. Hence

$$\pi_1(U(1)) = \pi_1(S^1) = \mathcal{Z}, \ \pi_n(U(1)) = \mathbf{I}, \ n > 1.$$

Next consider $SU(2)$ which can be represented by two complex numbers z_1, z_2 with the condition $|z_1|^2 + |z_2|^2 = 1$. This is because an arbitrary $SU(2)$ matrix is given by $\begin{pmatrix} z_1 & z_2 \\ -z_2^* & z_1^* \end{pmatrix}$ with determinant equals 1. Hence, it is easy to see the group has the structure of S^3. Hence

4.1 Point Defects, Line Defects, and More

$\pi_0(SU(2)) = \pi_1(SU(2)) = \pi_2(SU(2)) = \mathbf{I}$, $\pi_3(SU(2)) = \mathcal{Z}$. We can also obtain the homotopic groups of other compact semisimple groups like $SU(N)$. In fact $\pi_3(SU(N)) = \mathcal{Z}$, $N > 1$.

Given the above results, we can obtain the homotopy groups of the cosets $SU(2)/H$. These play an important role for the fundamental group of many physical order parameter spaces. Hence, the defects and phases of several materials can be studied. We will explain these with several examples now.

Consider a continuous group G and let H be the subgroup of G. We would be interested in general the coset space G/H. Consider a set of elements in H given by N such that these are connected to the identity by continuous paths in H itself. Then N is a normal subgroup of H and the fundamental group of G/H is given by H/N. i.e., $\pi_1(G/H) = H/N$. This result follows from results about exact sequences of groups and coset spaces (see Appendix at the end of this Chapter). But simple physical arguments can be used to explain this theorem.

1. Consider the Z_2 subgroup of $SU(2)$ given by $U \to -U$. The coset space is actually the group $SO(3)$. This can be seen from the result that $SO(3)$ is doubly connected and $SU(2)$ is its universal covering space. This will make $\pi_1(SO(3)) = Z_2$ as N is trivial in this case.
2. If the subgroup is continuous, for example, $U(1)$ of $SU(2)$. then the connected subset N is the whole of $U(1)$. Hence H/N is trivial and the fundamental group is also trivial. It is easy to see since $SU(2)/U(1)$ is known to be given by the manifold S^2 and $\pi_1(S^2) = I$.
3. If the subgroup consists of many disjoint components like for example $U(1) \times Z_2$, then the π_1 will be given by the set of disjoint elements. For example $\pi_1(SO(3)/U(1) \times Z_2) = Z_2$.

Basically, the subgroup maps the elements of the group to other points. A loop close to identity will remain a loop and can be continuously shrunk. But a path in the group space G linked by an element of H will become a loop in the coset space G/H. Still H consists of disjoint pieces obtained by N, then only those loops linked by H/N will be homotopically nontrivial.

Appendix: Exact Sequences

Consider map from a elements of a group $f : X \to Y$. The Image $Im(f)$ is the set of elements of Y which are mapped from the all elements of X. The Kernel of the map $Ker(f)$ is the set of elements of X which are mapped to the identity element of Y.

$$Im(f) : \{y \in Y \,|\, f(x) = y, \; for \; all \; x \; in \; X\};$$
$$Ker(f) : \{x \in X \,|\, f(x) = e\}$$

We can have a sequence of maps $X_1 \xrightarrow{f_1} X_2 \xrightarrow{f_2} X_3 \xrightarrow{f_3} X_4, \ldots$ Such a sequence is called exact if $Im(f_i) = Ker(f_{i+1})$. The exact sequences play an important role in classifying maps and finding homotopic invariants of topological spaces in particular group manifolds and their coset spaces.

When the sequence is finite with only spaces and ending with identity we get several invariants easily. Consider for example

$$0 \to X \to Y \tag{4.8}$$

Since it starts with 0 the sequence will be exact only if the second map $X \to Y$ is such that its Kernel is $\{0\}$. That makes the map one to one. Similarly consider the map

$$X \to Y \to 0 \tag{4.9}$$

Following the same arguments as above, the map $X \to$ must have the image as the whole of Y. That makes the map as onto. Now consider what is known as a short exact sequence

$$0 \to X \to Y \to 0 \tag{4.10}$$

Using the previous arguments the above will be an exact sequence only if $X \to Y$ is one-to-one and onto. That makes X as isomorphic to Y. A simple example can be given set of integers, even integers, and integers modulo 2. Consider

$$0 \to \mathcal{Z} \xrightarrow{2x} \mathcal{Z} \to \frac{\mathcal{Z}}{2\mathcal{Z}} \to 0 \tag{4.11}$$

This can be shown to be an exact sequence. This leads to the interesting result about the group, normal subgroup and coset space

$$e \to N \to G \to G/N \to e \tag{4.12}$$

Consider a compact semisimple group G, a subgroup H and the coset space G/H. We call G/H as the base manifold, H as the fiber over it and G as the bundle. Then an exact sequence follows from the maps about homotopy groups.

$$..\pi_{n+1}(G/H) \to \pi_n(H) \to \pi_n(G) \to \pi_n(G/H) \to \pi_{n-1}(H)\ldots \tag{4.13}$$

This powerful result can be exploited to obtain the homotopy invariants of several manifolds.

For example, the group $SU(2)$ is nothing but S^3 as mentioned earlier. It has a subgroup $U(1)$ which is S^1. The quotient $SU(2)/U(1) = S^2$. This is also known as Hopf fibration. Using our exact sequence we can now write:

$$\ldots \pi_3(U(1)) \to \pi_3(SU(2)) \to \pi_3(S^2) \to \pi_2(U(1)) \ldots \quad (4.14)$$

This gives

$$0 \to \mathcal{Z} \to \pi_3(S^2) \to 0 \quad (4.15)$$

Given the above short exact sequence, we can conclude

$$\pi_3(S^2) = \mathcal{Z} \quad (4.16)$$

It is interesting that similar exercise leads to

$$\pi_4(S^3) = \pi_4(S^2) = \mathcal{Z}_2 \quad (4.17)$$

A similar analysis leads to $SU(3)/SU(2) = S^5$ Using this we can understand $SU(3)$ as $S^5 \otimes_t S^3$, where subscript 't' indicates twisted product. We can again get a new exact sequence and obtain homotopic invariants.

$$\to \pi_4(S^5) \to \pi_3(SU(2)) \to \pi_3(SU(3)) \to \pi_3(S^5).. \quad (4.18)$$

This gives the sequence

$$0 \to \mathcal{Z} \to \pi_3(SU(3)) \to 0 \quad (4.19)$$

thereby giving isomorphism of $\pi_3(SU(3))$ and $\pi_3(SU(2))$ which is $\mathcal{Z}$. This result will be useful in instanton physics in Yang Mills theory. Following same procedure we can easily establish $[S^3, SU(N)]$ is $\mathcal{Z}$ for any N.

Exercises

4.1 Show

$$\to \pi_4(S^5) \to \pi_3(SU(2)) \to \pi_3(SU(3)) \to \pi_3(S^5). \quad (4.20)$$

Show this leads to $\pi_3(SU(3)) = \mathcal{Z}$.

4.2 Compute $\pi_2(RP^2)$ and therefore provide arguments to show that the defects in Nematic liquid crystals can be topologically stable.

4.3 Consider a two-dimensional medium, where the order parameter space is a vector field $\vec{V} = (a(x, y), b(x, y))$. Consider a curve C in the medium such that $\vec{V}$ is nonsingular on it. Show that the winding number N_V of $\vec{V}$ is given by

$$N_V = \frac{1}{2\pi} \oint_C \frac{adb - bda}{a^2 + b^2}$$

4.4 Show that $\frac{SO(3)}{SO(2)}$ is isomorphic to S^2. Generalize this for $\frac{SO(N+1)}{SO(N)}$, $N > 2$. Using the exact sequence, show that $\pi_4(S^3) = \pi_4(S^2) = \mathcal{Z}_2$.

References

1. C. Kittel, *Introduction to Solid State Physics* (John Wiley & Sons Publishers, 2015)
2. M. Nakahara, *Geometry, Topology and Physics* (IOP Publishing House, Bristol, 2003)
3. N.D. Mermin, The topological theory of defects in ordered media. Rev. Mod. Phys. **51**, 591 (1979)

Configuration Space Topology and Topological Conservation Laws

With formal notations on topological spaces, homotopy, homology and cohomology introduced in the first three chapters, we have taken up their applications in Chaps. 4–6. Particularly, in this chapter we will analyse how quantum physics in some simple systems will bring out new and additional features depending on such topological properties of space. In fact, the dynamical evolution of the system takes place in such topological spaces which are referred to as the *configuration space*.

5.1 Paths and Path Connectedness, Homotopy

In order to understand the effect of path connectedness of topological spaces on quantum systems, we will concentrate on the following smooth manifolds as configuration spaces:

1. Connected one-dimensional manifolds
   (a) Infinite in length like real line $\mathbb{R}^1$
   (b) Compact space like circle $\mathbf{S}^1$
   (c) Finite in length like closed interval $[-1, 1] \subset \mathbb{R}^1$
   (d) Semi-infinite in length like $\mathbb{R}_+ = [0, \infty)$
2. Two dimensional smooth manifold $\mathbb{R}^2 - \mathbf{D}^2$.
3. Three dimensional manifold $\mathbf{S}^3/\mathbb{Z}_2$
4. Besides these possibilities, there could also be disconnected pieces as well as curved topological spaces.

We will first focus on the quantum mechanical systems on some of these topological spaces. Then, we will also review quantum field theory on such topological spaces. These discussions will give an idea as to how topology plays a crucial role in obtaining energy spectrum, wave function and topological currents.

Recall, the first case 1.(a) is the familiar space taught in all quantum mechanics textbooks giving plane waves as wavefunctions

$$\Psi(x \in \mathbb{R}^1) = \exp\left(\frac{ipx}{\hbar}\right),$$

where p is the momentum of the free particle. The wave functions are constructed as superpositions of plane waves obeying square integrability criteria. Note that the position and momentum operators are unbounded in this case but need to be defined with proper domains in the other topological spaces. In the following subsection, we will explain formally the connection between quantum ambiguity in the wavefunction and the fundamental group of topological spaces.

5.1.1 Quantization Ambiguity and Fundamental Group

When quantising a physical system defined on a configuration space $\mathcal{Q}$, the standard procedure is to take the quantum mechanical state vector Ψ to be maps from $\mathcal{Q}$ to complex numbers $\mathbb{C}$:

$$\Psi : \mathcal{Q} \longrightarrow \mathbb{C}. \tag{5.1}$$

$|\Psi|^2$ provides us the probability density. In simply connected spaces it is expected Ψ to be a single-valued function. In spaces with multiple connectivity, one can allow multivalued wavefunctions since what is required is single-valued probability density. This provides the source of the ambiguity that Ψ can be multivalued up to a phase.

As the particle transverses a 'loop' ℓ and return to the starting point, the phase ambiguity of the wavefunction is provided by the homotopy class of the loop ℓ in that topological space:

$$\Psi(q) \xrightarrow{\ell} \Psi(q)\, e^{i\alpha[\ell]}.$$

Recall, the homotopy class refers to the set of loops which are deformable into one another. We will assume that there is no magnetic field in the region $\mathcal{Q}$.

We shall now establish that the allowed phases are fixed by the representations of fundamental group $\pi_1(\mathcal{Q})$. If we transverse the loop ℓ_1 followed by ℓ_2 the wavefunction will change to

$$\Psi(q) \xrightarrow{\ell_1} e^{i\alpha[\ell_1]}\Psi(q) \xrightarrow{\ell_2} e^{i\alpha[\ell_2]}e^{i\alpha[\ell_1]}\Psi(q).$$

Composing the two loops $\ell_1 \bullet \ell_2$ will correspond to ℓ_3 in $\mathcal{Q}$. Hence

$$\Psi(q) \xrightarrow{\ell_1 \bullet \ell_2 = \ell_3} \epsilon^{\alpha[\ell_3]}\, \Psi(q).$$

Hence, in the absence of magnetic field in Q, $\alpha[\ell]$ will form a one dimensional representation of $\pi_1(Q)$. This leads to the general theorem:

> **Theorem 5.1** *The quantisation ambiguities are characterised by the one-dimensional representations of the the fundamental group of the configuration space.*

We will illustrate the profound consequence of this theorem by studying quantum mechanical systems on the enumerated topological spaces (1.(b)- 3) in the following section.

5.2 Quantum Mechanical Systems on Different Topological Spaces

5.2.1 A Particle Moving on a Circle S^1

Consider a particle of mass m constrained to move on a circle $S^1 \equiv Q$. We will specify the position of the particle on a unit circle using angle ϕ. The Hamiltonian describing such a particle moving on a unit circle is

$$\hat{H} = \hat{p}_\phi^2 = -\frac{\partial^2}{\partial \phi^2}. \tag{5.2}$$

We have taken for simplicity $2m = 1$ and work with the units $\hbar = 1$. The time-dependent equation is

$$i\frac{\partial}{\partial t}\Psi = \hat{H}\Psi. \tag{5.3}$$

Normally we use periodic boundary conditions. In principle what is required as mentioned earlier is the probability density $P(\phi)$ to be single valued and it implies that should be periodic. Since $P(\phi) = |\Psi|^2$ we have the freedom to assign $\Psi(\phi + 2\pi) = e^{i\Theta}\Psi(\phi)$ where Θ is the arbitrary parameter. Solving Eq. (5.3) with the new requirement is straightforward exercise to give wavefunction and the corresponding energy eigenvalues:

$$\Psi_n(\phi, t) = \frac{1}{\sqrt{2\pi}} \exp(-iE_n t) \, \exp\left[i(n + \frac{\Theta}{2\pi})\phi\right],$$

$$E_n = \left(n + \frac{\Theta}{2\pi}\right)^2, \tag{5.4}$$

where $n \in \mathbb{Z}$. This simple periodicity property on circle $\mathbf{S}^1 \equiv \mathcal{Q}$ introduces Θ as the quantisation ambiguity parameter. We know that the configuration space is multiply connected and characterised by the first fundamental homotopy group

$$\pi_1(\mathbf{S}^1) = \mathcal{Z}.$$

The one-dimensional representations for this group are given by $g_n = e^{i\Theta n}$.

Even though we illustrated the phase ambiguity of wave function for $\mathcal{Q} = \mathbf{S}^1$, such ambiguity is generic for all configuration spaces $\mathcal{Q}$ with non-trivial first fundamental homotopy group. In other words, the ambiguity is characterised by the one-dimensional representations of $\pi_1(\mathcal{Q})$.

We can go back to the conventional periodic boundary condition through a unitary (though singular) transformation. Under the following transformation of the wave function:

$$\Psi'(\phi) = U \Psi(\phi) = \exp\left(-i\frac{\Theta\phi}{2\pi}\right) \Psi(\phi), \qquad (5.5)$$

we can see that the new wave function is periodic and hence single-valued:

$$\begin{aligned}
\Psi'(\phi + 2\pi) &= \exp\left(-i\frac{\Theta\phi}{2\pi} - i\Theta\right) \Psi(\phi + 2\pi) \\
&= \exp\left(-i\frac{\Theta\phi}{2\pi} - i\Theta\right) e^{i\Theta} \Psi(\phi) \\
&= \exp\left(-i\frac{\Theta\phi}{2\pi}\right) \Psi(\phi) = \Psi'(\phi).
\end{aligned}$$

The Hamiltonian under such a transformation will become

$$\hat{H}' = U^\dagger \hat{H} U = \exp\left(i\frac{\Theta\phi}{2\pi}\right) \hat{H} \exp\left(-i\frac{\Theta\phi}{2\pi}\right).$$

Hence the new Hamiltonian is:

$$H' = \frac{1}{2}\left(p_\phi + \frac{\Theta}{2\pi}\right)^2. \qquad (5.6)$$

Usually, the Lagrangian dynamics are useful to illustrate the paths in the configuration space $\mathcal{Q}$. Hence we would like to understand the effects of the above transformations within the Lagrangian framework. The new Lagrangian is given by

$$L' = \frac{1}{2}(\dot{\phi})^2 - \frac{\Theta\dot{\phi}}{2\pi}. \qquad (5.7)$$

5.2 Quantum Mechanical Systems on Different Topological Spaces

The additional term is a total time derivative and formally does not change Euler-Lagrange equations. Note that the new action is no longer time reversal invariant for $\Theta \neq 0, \pi$. In other words, $S'(\Theta) \xrightarrow{t \to -t} S'(\Theta) + 2\Theta$ implying that the partition functional $Z = e^{iS'}$ is time reversal invariant when $\Theta = 0$ or π.

Incidentally, the Θ for this simplest system resembles the Θ vacua in quantum chromodynamics (QCD) which is a non-abelian gauge theory [1,2]. Similar to the addition of an extra degree of freedom through axion field [3] which controls QCD Θ vacua, it is straightforward to make the ground state energy independent of Θ (see exercise 5.5).

With this elaborate discussion for $\mathcal{Q} = \mathbf{S}^1$, we will compare and contrast with the particle moving on an interval in the following subsection.

5.2.2 Particle Confined Within the Interval [−1, 1]

Topologically, the interval refers to a manifold with a boundary. Quantum mechanics of a particle within the interval is the well-known 'particle in a box' which all undergraduates have been taught. Unlike the standard textbooks with the wave function at the boundaries obeying $\psi(1) = \psi(-1) = 0$, we could work with

$$\psi(-1) = \exp(i\Theta)\psi(1),$$

which resembles $\mathcal{Q} = \mathbf{S}^1$ for $\Theta \neq 0$ or more complicated boundary conditions. Even the choice of wavefunction vanishing at the boundary, we can have situations where the action of the observables like Hamiltonian can project the function to a new one which does not vanish at the boundary. For example, consider

$$\psi(x) = N(x-1)(x+1),$$

vanishing at the boundaries $x = \pm 1$ which is not an eigenstate. The action of Hamiltonian on this wavefunction

$$\hat{H}\psi(x) = -\frac{d^2}{dx^2}\psi(x) = -2N$$

gives a new function not in the domain of $\hat{H}$. But the correct way of understanding this is by expanding this wavefunction in the basis of eigenstates of the Hamiltonian and act on the expansion.

From the perspective of self-adjoint operators discussed in Chap. 3, the question which will be of interest is whether momentum $\hat{p} = -i\frac{d}{dx}$ or Hamiltonian $\hat{H} = -\frac{d^2}{dx^2}$ is a self adjoint operator with the conditions imposed on the wave function at the boundary. The answer to this question depends on the boundary conditions for wave functions $\psi(-1), \psi(1)$ as well as derivative of the wave functions $\psi'(-1), \psi'(1)$. Recall the discussion in Chap. 3 where we had established that the Hamiltonian operator has deficiency indices (2,2) and has the ambiguity of 4 parameters. The Hamiltonian admits self-adjoint extension in the domain of functions such that

$$\begin{pmatrix} \psi(-1) \\ \psi'(-1) \end{pmatrix} = U \begin{pmatrix} \psi(1) \\ \psi'(1) \end{pmatrix} \tag{5.8}$$

where U is an arbitrary unitary matrix. Arbitrary unitary matrix has 4 parameters and hence the quantization of this Hamiltonian is subjected to this ambiguity. The various options for wave functions with different boundary conditions provide examples where there will be differences in the quantum theory. Through the exercise 5.6, the readers can understand the role of domains of self-adjoint operator and their extensions for the wave function $\psi(x) = N(x-1)(x+1)$. We will now look at the quantum features for the positive real line $\mathbb{R}^1_+$.

5.2.3 Particle on $\mathbb{R}^1_+$

The manifold $\mathbb{R}^1_+$ has a boundary at $x = 0$. From the perspective of self-adjoint operators and the deficiency index discussed in Chap. 3, the Hamiltonian allows the following one parameter space of boundary conditions:

$$\psi'(0) = \kappa \psi(0), \tag{5.9}$$

where the parameter κ interpolates between Dirichlet ($\psi(0) = 0$, $\kappa = \infty$) and Neumann conditions($\psi'(0) = 0$, $\kappa = 0$). Interestingly there is a bound state localised close to the boundary for different conditions. The energy of the bound state is given by $-\kappa^2$. Students may compare with the well-known case of quantum mechanics of particles on $\mathbb{R}^1$ with a delta function potential at the origin. Interestingly momentum operator on $\mathbb{R}^1_+$ is not self-adjoint since the deficiency indices turn out to be $(1, 0)$. We can think of the delta function potential as the union of $\mathbb{R}^1_\pm$ with deficiency indices as $(1, 1)$ and hence can be extended with suitable boundary conditions at the origin.

So far, we confined to one-dimensional topological spaces. We will address the energy spectrum for the particle moving in the two-dimensional topological spaces in the following subsection.

5.2.4 Particle Moving in the Space $\mathbb{R}^2 - \mathbf{D}^2$

The two-dimensional real space $\mathbb{R}^2$ is simply connected. If we remove the disc $\mathbf{D}^2$ from $\mathbb{R}^2$, we get the topological space $\mathbb{R}^2 - \mathbf{D}^2$ with a boundary circle $\mathbf{S}^1$ of radius R. This example is interesting because it introduces novelty to von Neumann's theorem of deficiency spaces. The Hamiltonian for a free particle in such a space is $-\nabla^2$ which has deficiency indices (∞, ∞). Even though there are an infinite number

5.2 Quantum Mechanical Systems on Different Topological Spaces

of boundary conditions, those obeying rotational symmetry are few. One important boundary condition in this context is

$$\kappa \Psi(R) + \partial_r \Psi(r)|_{r=R} = 0, \qquad (5.10)$$

where the parameter κ interpolates between Dirichlet and Neumann conditions here as well. From the above condition, we can see κ has a dimension of inverse length. This boundary condition is known in the literature as Robin boundary condition [4]. A self-adjointness analysis of the Hamiltonian shows that it is essentially self-adjoint when Robin's boundary conditions (5.10) are satisfied [5].

The solution for the Schrodinger equation (eigenvalue equation)

$$\nabla^2 \Psi(r, \theta) = -\lambda \Psi(r, \theta) \qquad (5.11)$$

is easy to obtain:

$$\Psi(r, \theta) \propto K_n(\sqrt{\lambda} r) e^{in\theta} \qquad (5.12)$$

where K_n are the Bessel functions of the third kind and $\lambda \geq 0$ is the bound state eigenvalue. Note that the Hamiltonian $-\nabla^2$ on $\mathbb{R}^2$ is a positive definite operator without any bound state. Interestingly, the number of bound state solutions N localised on the boundary is given by:

$$N \propto \kappa R \qquad (5.13)$$

Hence N is proportional to the radius R of the circular boundary. So far, we have addressed the crucial role played by the boundary condition on a particle moving on topological spaces with boundary. In the following subsection, we will review rigid rotor which will be useful for understanding spin statistics theorem to be discussed in the subsequent chapter.

5.2.5 Rigid Rotor

The rigid rotor is any rigid body with its center of mass fixed. It is kinematically characterised by the moment of Inertia tensor I. We will assume that body is spherical so that it it possesses rotational symmetry. The Hamiltonian for a rigid rotor is well known from classical mechanics as:

$$H = \frac{\vec{L}.\vec{L}}{2I}$$

where $\vec{L}$ is the angular momentum operator. The freedom for the rotor comes from all possible rotations one can perform keeping the center of mass fixed. For the spherically symmetric rigid body, we know that the space of rotations is characterised by 3×3 orthogonal matrix with determinant 1. Hence the configuration space describing the rigid rotor is $\mathcal{Q} = SO(3)$. The space is homeomorphic to $\frac{SU(2)}{\mathbb{Z}_2}$. Since the $SU(2) \cong S^3$ and is simply connected, the fundamental group of $\mathcal{Q} = \frac{SU(2)}{\mathbb{Z}_2}$ is $\mathbb{Z}_2$. There are two one dimensional representations for this configuration space $\mathcal{Q} = \frac{SU(2)}{\mathbb{Z}_2}$:

- trivial representation
- $1, -1$.

The wave functions describing rigid rotor belong to the Hilbert space $\mathcal{H}$ which splits into two classes:

$$\mathcal{H}_{even} = \sum_{j=\mathbb{Z}^+} \mathcal{D}^j \; ; \; \mathcal{H}_{odd} = \sum_{j=\mathbb{Z}^+ + \frac{1}{2}} \mathcal{D}^j \qquad (5.14)$$

The integer j representations are such that the wave function returns to its original value when a rotation by angle 2π is performed about any axis. But in the half-odd integral representation, the wave function gets multiplied by a minus sign ($e^{i\pi} = -1$) when a full rotation by 2π is made. Only rotation by 4π returns the wave function of arbitrary j to its original value. Incidentally, the group manifold of $SO(3)$ can be represented by a solid sphere whose radius ranges from $0 \leq r \leq \pi$ with the diametrically opposite points on the boundary two-sphere $\mathbb{S}^2$ of radius $r = \pi$ identified. Hence the topological space $\mathcal{Q} = SO(3)$ is doubly connected. So, we get two inequivalent quantisations corresponding to integer and half-odd integer spin j representations. This naturally corresponds to the two types of particles obeying familiar statistics (Bose-Einstein distribution for bosons and Fermi-Dirac distribution for fermions) which we study in statistical mechanics. Till now, we discussed the impact of the topological space on the wave function and energy spectrum of a quantum mechanical system. In the following section, we will consider the configuration space $\mathcal{Q}$ for N identical particles moving in d-dimensional space $\mathbb{R}^d$ and the emergence of statistics from topological arguments.

5.3 System of N Identical Particles and Origin of Statistics

The configuration space describing a system of N identical particles in $\mathbb{R}^d$ is $\mathcal{Q} = (\mathbb{R}^d)^N$. If we treat these particles as indistinguishable, then the action of any element of permutation group S_N ($\mathcal{P} \in S_N$) on these particles in the configuration $\mathcal{Q} = (\mathbb{R}^d)^N$ located at different spatial points ($\vec{r}_1 \neq \vec{r}_2 \neq \ldots \neq \vec{r}_N$) will permute the location of the particles giving a different point in $\mathcal{Q}$:

$$(\vec{r}_1, \vec{r}_2, \ldots \vec{r}_N) \to \mathcal{P}(\vec{r}_1, \vec{r}_2 \ldots \vec{r}_N) . \qquad (5.15)$$

5.3 System of N Identical Particles and Origin of Statistics

These two distinct points in $Q = (\mathbb{R}^d)^N$ need to be identified as the particles are indistinguishable. Further, some permutation elements of the permutation group will leave the points in $Q = (\mathbb{R}^d)^N$ unchanged when two or more particles are at the same location ($\vec{r}_i = \vec{r}_j$ for some $i \neq j$). These invariant points in Q are referred to as fixed points under S_N group operation. In other words, S_N group action on Q is not a free action. Hence, the quotient space $(\mathbb{R}^d)^N/S_N$ is a singular space because of the fixed points. In order to make the quotient space a smooth manifold, we need to remove such fixed points from Q before applying the permutation operation $\mathcal{P} \in S_N$. Technically this can be achieved by imposing hard-core repulsion amongst these indistinguishable particles. Hence the configuration space describing N indistinguishable particles with hard core repulsion will be

$$Q = \frac{(R^d)^N - \{Diagonals\}}{S_N}. \quad (5.16)$$

We have indicated the hard-core condition by removing those points where two or more particles are at the same location. We have called them as 'Diagonals'. What is the fundamental group $\pi_1(Q)$? We mentioned earlier if the numerator space is simply connected then the fundamental group is the discrete group which is used to quotient the numerator space. In this case, it is easy to see that the fundamental group of $(R^d)^N - \{Diagonals\}$ is trivial as long as $d \geq 3$. This is because removing points in dimensions greater than 2 will not create obstructions for loops to deform arbitrarily [6]. The loops can always be moved far away and contracted to the identity loop. However, the loops cannot be deformed to the identity loop in the two-dimensional space ($d = 2$). Hence the fundamental group is

$$\pi_1(Q) = S_N \text{ for } d \geq 3. \quad (5.17)$$

We are interested in the common one-dimensional representations of S_N which appear for any N. There are two of them.

1. The trivial representation which does not introduce a phase when we do any permutation of particles. That is a symmetric wave function:

$$\Psi(r_1, r_2, ..r_i, ...r_j...r_N) = \Psi(r_1, r_2, ...r_j, ...r_i, ...r_N), \quad (5.18)$$

   for all $i \neq j \in (1, 2, ... N)$. This results in particles in nature called 'bosons'.

2. The second representation divides the permutation into odd and even permutations. Odd ones are those which are obtained by an odd number of transpositions or exchanges. The even permutation refers to even the number of exchanges or transpositions. The second representation is such that the wave function picks up ± 1 under permutation operation:

$$\Psi(r_1, r_2, ..r_i, ...r_j....r_N) = (-1)^n \Psi(r_1, r_2, ...r_j, ...r_i, ...r_N), \quad (5.19)$$

where n refers to the number of transpositions or exchanges in the permutation. Hence the wave function picks up a $\pm$ sign depending on the nature of the permutation:

$$\text{Even Permutation} = 1, \text{ Odd Permutation} = -1$$

The wave function with the property (5.19) refers to particles called 'fermions'. Thus our discussion on the one-dimensional representation of the fundamental group $\pi_1(\mathcal{Q})$ for $\mathcal{Q}$ (5.16) reproduces the familiar bosonic (totally symmetric) and fermionic (totally antisymmetric) wave function studied in quantum statistical mechanics [7].

We have to address the fundamental group of $\mathcal{Q}$ for $d = 2$ which is the theme of the following subsection leading to new type of particles called 'anyons'.

5.3.1 Anyons in $d = 2$ Dimensions

What we established for statistics work well for dimensions $d \geq 3$. This is because in two dimensions removal of points create holes in the space affecting fundamental representation. Hence the fundamental group of numerator space (5.16) is more complicated. Interestingly the fundamental group of $\mathcal{Q}$ for $d = 2$ is known as the 'braid group' with N strands denoted by $\mathcal{B}_N$. We refer the readers to the article [4] to see why the wave function of identical particles in two dimensions under exchange picks up an arbitrary phase and not ± 1 sign:

$$\Psi(r_1, r_2, \ldots r_i, \ldots r_j, \ldots r_N) = e^{i\Theta} \Psi(r_1, r_2, \ldots r_j, \ldots r_i, \ldots r_N) . \tag{5.20}$$

These exchange operations of identical particles on a two-dimensional plane can be viewed as a clockwise exchange which is distinct from the corresponding anticlockwise exchange. In fact, these exchange operations can be viewed as over-crossing (for clockwise exchange) and under-crossing (for anticlockwise exchange) amongst world lines of such particles. This leads naturally to the following presentation of braid groups $\mathcal{B}_N$:

Consider two rods and N strings (world lines of identical particles) from bottom to top (fixed initial time t_i to fixed final time t_f). The clockwise (anticlockwise) exchange of particle i and particle $i + 1$ will be denoted as over crossing b_i (under crossing b_i^{-1}). In fact, the elements b_i with $i = 1, 2, \ldots N - 1$ are the generators of the braid group $\mathcal{B}_N$. Any permutation of identical particles can be written in terms of these words constructed using $b_i^{\pm}$ where $i \in (1, N)$. It is easy to see the following defining relations amongst the generators:

$$b_i b_j = b_j b_i, \quad |i - j| \geq 2 . \tag{5.21}$$

The following example belongs to braid group $\mathcal{B}_6$ illustrates the above relation (Fig. 5.1).

5.3 System of N Identical Particles and Origin of Statistics

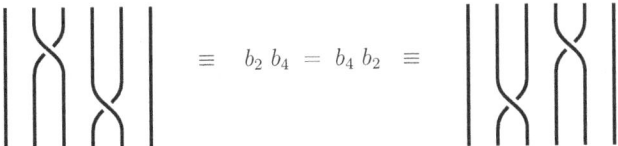

Fig. 5.1 Braiding relation 1

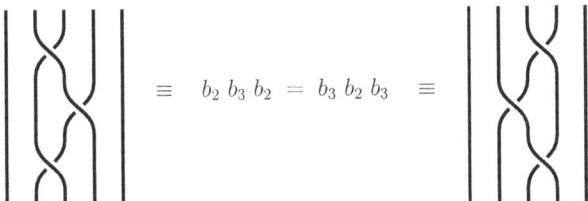

Fig. 5.2 Braiding relation 2

Further, for braiding involving i-th strand and $i + 1$-th strand, the following relation

$$b_i \, b_{i+1} \, b_i = b_{i+1} \, b_i \, b_{i+1} \,, \tag{5.22}$$

is satisfied. For the braid diagram belonging to $\mathcal{B}_6$, it is easy to pictorially verify (Fig. 5.2).

Clearly, the number of elements belonging to $\mathcal{B}_N$ is infinite even for $N = 2$.

Note We will get the permutation group S_N if we impose an additional condition:

$$b_i^2 = 1, \; \forall \, i \,. \tag{5.23}$$

It is this condition (5.23) that leads to the difference between $\mathcal{B}_N$ and S_N and in turn, leads to the distinction of statistics of particles in two and three dimensions. It is well known that one-dimensional representations of the braid group with an arbitrary number of braids are characterised by

$$b_i = e^{i\Theta}, \; \forall \, i \,. \tag{5.24}$$

These correspond to *fractional statistics* and the wave functions of the particles under exchange obey Eq. (5.20). These particles are known as 'anyons' where $\Theta \in (0, \pi)$. Recall, that such a phase Θ is possible only in two dimensions as clockwise exchange is distinct from anticlockwise exchange. When $\Theta = 0$ we get bosons and $\Theta = \pi$ we obtain fermions. The phase Θ of anyons are interpolating between bosons ($\Theta = 0$) and fermions ($\Theta = \pi$).

Even though we elaborated braids on $\mathbb{R}^2$, we could also consider braids on compact two-dimensional surfaces like sphere $\mathbf{S}^2$ and torus $\mathbf{T}^2$ or two-dimensional surfaces with arbitrary genus 'g'. The study of representations of braid group will be taken up in the following chapter.

So far, our discussions have been on quantum mechanical systems dependent on the topology of the configuration space. We will now take up some field theory examples in the following section.

5.4 Field Theory on Topological Spaces

We will study solitons and $O(3)$ non-linear sigma models to highlight the role of topology [8] in field-theoretic systems.

5.4.1 Kink Soliton

Consider a scalar field ϕ in $1+1$ dimension. We will see that the configuration spaces are not path connected, i.e. $\Pi_0(\mathcal{Q}) \neq 1$. Hence an interesting field theory model to study. The Lagrangian density describing the scalar field is

$$\mathcal{L} = \frac{1}{2}\left[\partial_\mu \phi \partial^\mu \phi - \lambda(\phi^2 - 1)^2\right]. \tag{5.25}$$

We are interested in static or time independent configurations of the classical equations of motion. They are obtained by minimising the time independent Hamiltonian:

$$\mathcal{H} = \frac{1}{2}(\nabla \phi)^2 + \lambda(\phi^2 - 1)^2 \tag{5.26}$$

To get finite energy, the configurations $\phi \to \pm 1$ as $x \to \pm \infty$. Otherwise the integral will diverge. The configuration space is given by the following:

1. $x \to -\infty, \phi \to 1, \ x \to +\infty, \phi \to 1$
2. $x \to -\infty, \phi \to -1, \ x \to +\infty, \phi \to 1$
3. $x \to -\infty, \phi \to 1, \ x \to +\infty, \phi \to -1$
4. $x \to -\infty, \phi \to -1, \ x \to +\infty, \phi \to -1$

These four spaces of configurations are disconnected and there is no path to go from one to other without affecting the behaviour at infinity. In other words there is infinite potential barrier between these classes of configurations. That means a configuration in one sector will remain always in that sector through time evolution preserving a number known as a soliton number. The solutions are

5.4 Field Theory on Topological Spaces

known as 'soliton' solutions. Remember the first and fourth can be homotopic to a configuration which is space-independent and hence such a solution will have zero energy without affecting boundary conditions. The rest of the cases with non-zero soliton numbers will necessarily have space dependence and will have non-zero energy. If $\mathcal{Q}$ is the configuration space $\pi_0(\mathcal{Q}) = Z_4$. Conservation of soliton number is not related to any symmetry and hence Noether's theorem will not be applicable. They are known as *topological* conservation laws. They are related to the conservation of a topological current which do not arise from any symmetries of the equations of motion, but only from boundary conditions. The topological current in this case is

$$j^\mu = \epsilon^{\mu\nu}\partial_\nu\phi. \tag{5.27}$$

It is easy to verify $\partial_\mu j^\mu = 0$. The conserved topological charge corresponding to the integral of the zeroth component is

$$Q = \int j^0\, dx = \int \partial_x\phi\, dx = \phi(+\infty) - \phi(-\infty) = \pm 2, 0. \tag{5.28}$$

The soliton solutions obeying $\phi(\pm\infty) = \pm 1$, is called kink and the ones obeying $\phi(\mp\infty) = \pm 1$ is known as anti-kink soliton in the literature(see [9]). We can obtain equations for static configurations by demanding $\frac{d}{dt}\phi(x) = 0$ in the Euler Lagrange equations:

$$\partial^2\phi = 4\lambda\phi(\phi^2 - 1) \tag{5.29}$$

Since we require $\phi \to \pm 1$ as $x \to \pm\infty$ we can start with an ansatz $\phi(x) = \tanh(cx)$. This on substitution in the equation for static configuration gives:

$$c^2 = 2\lambda. \tag{5.30}$$

Hence the static configuration in the topological sector is

$$\phi(x) = \tanh(\sqrt{2\lambda}x), \tag{5.31}$$

which we have plotted for the choice $2\lambda = 1$ below (Fig. 5.3).

We can substitute this static configuration in the Hamiltonian density $\mathcal{H}$ (5.26) and get the profile of the energy density.

$$\mathcal{H} = \frac{5\lambda}{4}\operatorname{sech}^4(\sqrt{2\lambda}x) \tag{5.32}$$

Fig. 5.3 Soliton number one solution

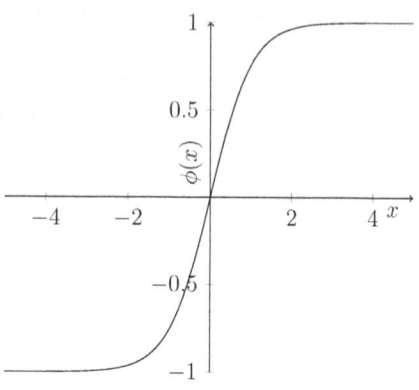

Fig. 5.4 Energy density of the soliton

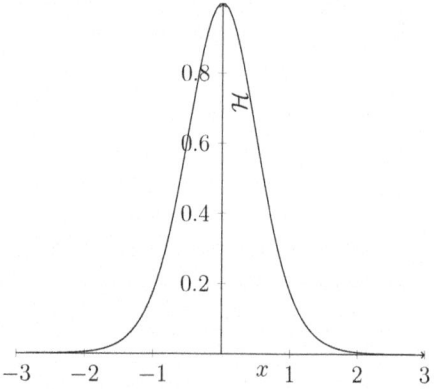

We can plot this profile (again with $2\lambda = 1$) which has a maximum at the origin and localised close to it as shown in Fig. 5.4.

The topological nature of the soliton solution is the underlying reason for the soliton to behave like an extended particle (a particle with size) under scattering. The Lagrangian and the Euler-Lagrange equations are symmetric under uniform motion. Hence we expect the configuration obtained by boost transformation is also a solution. It is easy to show

$$\phi = \tanh c(x - vt)$$

is also a solution for a time-dependent equation. Problem 5.10 in Exercises will be useful to deduce energy as a function of the velocity v.

5.4.2 O(3) Nonlinear σ Model

We will now study this theory in 2 + 1 dimensions. Here $O(3)$ indicates the symmetry group of the nonlinear σ model [10–12]. The field has three components ϕ_i with one condition $\sum \phi_i^2 = 1$. At every point in $\mathbb{R}^2$ the field maps a point on to a sphere $\mathbf{S}^2$. Such a configuration evolves in time so that the configuration $\phi_i(x, y, t)$ with $\sum_{i=1}^{3} \phi_i^2(x, y, t) = 1$. The Lagrangian density can be taken as:

$$\mathcal{L} = \frac{1}{2} \partial_\mu \vec{\phi} \cdot \partial^\mu \vec{\phi} . \tag{5.33}$$

We can impose the condition on the fields through Lagrangian multiplier λ and write the action functional as:

$$S[\phi] = \int d^3 x \left(\frac{1}{2} \partial_\mu \vec{\phi} \cdot \partial^\mu \vec{\phi} + \lambda(\vec{\phi} \cdot \vec{\phi} - 1) \right) . \tag{5.34}$$

The time independent equations of motion can easily be derived.

$$\nabla^2 \vec{\phi} - (\vec{\phi} \cdot \nabla^2 \vec{\phi}) \vec{\phi} = 0 . \tag{5.35}$$

The Hamiltonian for time independent solutions is easily seen to be:

$$H = \int d^2 x \, \frac{1}{2} \left(\partial_i \vec{\phi} \cdot \partial_i \vec{\phi} \right) . \tag{5.36}$$

Again we explore stationary configuration (time independent). They should be such that

$$\phi_i \xrightarrow{r \to \infty} \text{constant vector.}$$

This is needed to make the energy density in Eq. (5.36) go to zero sufficiently as $r \to \infty$. The total energy will be finite.

This enables us to identify all the points at ∞ in $\mathbb{R}^2$ as one point. We get a sphere $\mathbf{S}^2$. This process is known as one-point compactification.

Hence the configuration space consists of all possible functions from $\mathbf{S}^2 \to \mathbf{S}^2$. Such a collection of maps as pointed out earlier, in the Part 1 chapters, provides homotopic classes. Two maps in different classes are not path-connected. That means time evolution from one map will not reach another in a different class. This set of classes is distinguished by an integer from $\pi_2(\mathbf{S}^2) = \mathcal{Z}$. This implies the $\Pi_0(\mathcal{Q}) = \mathcal{Z}$. This follows from the result in the previous chapter $\Pi_0[\mathbf{S}^2, \mathbf{S}^2] = \Pi_2(\mathbf{S}^2) = \mathcal{Z}$. This number will be called the 'soliton number' as in the previous example.

It is natural to raise a question: Is there a topological conserved current as described earlier? The answer is yes. Consider the current

$$j^\mu = \epsilon^{\mu\nu\rho} \epsilon^{ijk} \phi_i \partial_\nu \phi_j \partial_\rho \phi_k . \tag{5.37}$$

We can easily prove this current is conserved. Since $\sum \phi_i^2 = 1$, we get

$$\sum \partial_\mu \phi_i \, \phi_i = 0 . \tag{5.38}$$

Hence the matrix

$$M_{\mu i} = \partial_\mu \phi_i$$

has a null vector ϕ_i. That means the matrix $\partial_\mu \phi_i$ is singular, i.e., $Det\, M_{\mu i} = 0$.

$$Det\, M = \partial_\mu \left(\epsilon^{\mu\nu\rho} \epsilon^{ijk} \phi_i \, \partial_\nu \phi_j \, \partial_\rho \phi_k \right) = 0 . \tag{5.39}$$

Hence $\partial_\mu j^\mu = 0$. This establishes the conservation of the topological current j_μ. What the charge of the current represents is equally an interesting question. What is the value of the charge obtained by integrating the zeroth component of the current j_μ? Consider

$$Q = \frac{1}{4\pi} \int j_0 \, d^2x = \frac{1}{4\pi} \int \epsilon^{lm} \epsilon^{ijk} \phi_i \partial_l \phi_j \partial_m \phi_k = N .$$

It is the winding number of maps from $\mathbf{S}^2 \to \mathbf{S}^2$. For example, consider maps of coordinates $\theta, \phi \to \Theta, \Phi$ such that $\Theta(\theta) = \theta, \Phi = N\phi$. The area of the unit sphere $\mathbf{S}^2$ is

$$\int \sin\Theta \, d\Theta \, d\Phi = N \int \sin\theta \, d\theta \, d\phi .$$

These different configurations provide different soliton numbers and will not change under time evolution whatever be the equation of motion.

As remarked earlier they do not correspond to symmetries but the topological characterization of the configuration space $\mathcal{Q}$. We can construct equations of motion for the $O(3)$ nonlinear sigma model and consider static solutions. The field equations are

$$\partial_\mu \partial^\mu \phi_i = 0, \text{ and } \phi_1^2 + \phi_2^2 + \phi_3^2 = 1 . \tag{5.40}$$

We can look for $\phi_i \xrightarrow{r \to \infty} (0, 0, 1)$ After imposing the constraint through the Lagrange multiplier we get the equation as:

$$\nabla^2 \phi - (\phi \cdot \nabla^2 \phi)\phi = 0. \tag{5.41}$$

This equation has nonsingular solutions classified by the homotopic classes. To achieve this, we can project the sphere onto the equatorial plane using complex field W:

$$W = \frac{2(\phi_1 + i\phi_2)}{1 - \phi_3}. \tag{5.42}$$

Then our static solutions surprisingly should satisfy the condition

$$\partial_{z^*} W = 0, \text{ where } z = x + iy. \tag{5.43}$$

This makes W a holomorphic function and solutions in the homotopic class with the soliton number 'N' can be easily written. They are given by

$$W = \left(\frac{z - z_0}{\lambda}\right)^N. \tag{5.44}$$

We can also look for fundamental group in the set of configurations of soliton number N. That is, we look for $\pi_1(\mathcal{Q})$. That is the $\pi_1[\mathbf{S}^2 \to \mathbf{S}^2]$. There is a simple result that is equal to $\pi_3(\mathbf{S}^2)$. It is also known that this is $\mathcal{Z}$. This invariant is known as the Hopf invariant. This plays an important role in determining the statistics of the solitons which will be clear from solving some of the problems in Exercises. The connections to the spin-statistics theorem will be presented in the subsequent chapter. The reader is referred to see the book by Rajaraman [9] for detailed discussions on solitons. In addition, we have provided a list of papers as references which may be useful for readers planning to pursue research in this area.

Exercises

5.1 Show the one-dimensional representations for the group $\mathcal{Z}$ are given by $g_n = e^{i\Theta n}$.

5.2 Obtain energy for a 'free' particle moving on $\mathbb{R}_+ = [0, \infty)$ and explain the role of topology. Compare with particle in $\mathbb{R}$ with $\delta(x)$ potential at the origin.

5.3 Discuss the delta function potential in the previous problem as quantum mechanics of a particle on a space which is the union of $\mathbb{R}_+ = [0, \infty)$ and $\mathbb{R}_- = [0, -\infty)$. Is momentum self adjoint operator in this example?

5.4 A particle is moving on two circles touching each other. Find the fundamental group of the configuration space and discuss quantization.

5.5 Consider a particle on a circle coupled to an oscillator given by the Lagrangian

$$L = \frac{1}{2}\dot{\theta}^2 + \frac{1}{2}\dot{x}^2 - \frac{1}{2}x^2 + gx\dot{\theta},$$

where g is a coupling constant. Show that Hamiltonian is independent of the inequivalent quantization parameter Θ, and the states correspond to shifted oscillator with the parameter Θ.

5.6 Solve QM of a particle in a one dimensional box with periodic boundary conditions. Modify the boundary conditions to introduce periodic up to a phase. Expand new eigenfunctions in terms of the previous eigenfunctions.

5.7 In the above problem with wavefunction vanishing at the boundary, namely $\psi(\pm 1) = 0$ consider a wavefunction $\psi(x) = N(x^2 - 1)$. Obtain the expectation value

$$\langle H^2 \rangle = \left\langle \psi \left| \left(-\frac{d^2}{dx^2} \right)^2 \right| \psi \right\rangle$$

Is this zero? Is it consistent? Get the correct value and explain the role of domain of the operator H.

5.8 Obtain the Hamiltonian density $\mathcal{H}$ for the soliton solution. Plot the energy density as a function of x. Show it is peaked around $x = 0$.

5.9 The topological current in Eq. (5.27) is $j^\mu = \epsilon^{\mu\nu} \partial_\nu \phi$. Verify $\partial_\mu j^\mu = 0$. Show that the conserved topological charge corresponding to the integral of the zeroth component is

$$\int j^0 \, dx = \pm 2, 0. \qquad (5.45)$$

5.10 Show $\phi = \tanh c(x - vt)$ is also a solution for time dependent equation. Provide arguments why this should be a solution. Obtain energy. Show it represents soliton moving with uniform velocity. Find the energy density. Discuss relativistic covariance.

5.11 Consider the Lagrangian density for a scalar field in 1+1 dimension given by

$$\mathcal{L} = \frac{1}{2}(\partial_\mu \phi)(\partial^\mu \phi) - (1 - \cos\phi). \qquad (5.46)$$

Discuss the topological solitons of the equations of motion. Provide an example of a kink/antikink soliton.

5.12 Derive the static (independent of time) equation of motion given in Eq. (5.35).

5.13 Provide an example of configuration with soliton number ± 1 for the non-linear sigma model in 2+1 dimensions. (Hint: A typical soliton configuration can be taken as: $\vec{\sigma} \cdot \vec{\phi} = \sigma^3 \cos f(r) + \hat{x} \cdot \vec{\sigma} \sin f(r)$ where $f(r) = \pi, f(\infty) = 0$.)

5.14 Map the problem to $\mathbf{CP}^1$ dynamics. (Hint: Define two complex variables $Z_i, i = 1, 2$, with $|Z_1|^2 + |Z_2|^2 = 1$. Define $\phi^a = \bar{Z}\sigma^a Z$. Show this satisfies $\sum (\phi^a)^2 = 1$. Rewrite the nonlinear sigma model in terms of the new fields Z_i.) Obtain the conserved current in the new field variables.

5.15 Obtain the Hopf invariant for the previous example of $\mathbf{CP}^1$ field in 2+1 dimensions. Obtain angular momentum operator.

5.16 Suppose the configuration space is $\mathcal{Q} = SO(3)$. Show that the fundamental group of $\mathcal{Q}$ is $\mathbb{Z}_2$. What are the one-dimensional representations.

5.17 Show the fundamental group of $\mathbf{RP}^2$ is $\mathbb{Z}_2$.

5.18 Show that braid group with restriction $b_i^2 = 1$ is the same as the permutation group. Provide examples of one and two-dimensional representations.

References

1. G. t'hooft, Phys. Rev. Lett. **37**, 8 (1976)
2. R. Jackiw, C. Rebbi, Phys. Rev. Lett. **37**, 176 (1976)
3. R. Peccei, H. Quinn, CP conservation in the presence of pseudoparticles. Phys. Rev. Lett. **38**(25), 1440 (1977)
4. T.R. Govindarajan, Rakesh Tibrewala. Phys. Rev. D **83**, 124045 (2011)
5. M. Asorey, A. Ibort, G. Marmo, Int. J. Geom. Methods Mod. Phys. **12**, 1561007 (2015)
6. J.M. Leinaas, J. Myrheim, Nuovo Cimento **B37**, 1 (1977)
7. P. Ramadevi, Exchange of identical particles. Reson. J. 23–28 (Feb 2001)
8. D. Finkelstein, J. Math. Phys. **7**, 1218 (1966)
9. R. Rajaraman, *Kink Solution: Solitons and Instantons: An Introduction to Solitons and Instantons in Quantum Field Theory* (North-Holland, Amsterdam, 1982)
10. F. Wilczek, Phys. Rev. Lett. **48**, 1144 (1982); ibid., 1146 (1982); ibid. **49**, 957 (1982); F. Wilczek, A. Zee, Phys. Rev. Lett. **51** (1983)
11. A.P. Balachandran, G. Marmo, A. Stern, B. Skagerstam, *Classical Topology And Quantum States* (World Scientific, Singapore, 1991)
12. T.R. Govindarajan, R. Shankar, N. Shaji, M. Sivakumar, Phys. Rev. Lett. **69**, 721 (1992)

Spin-Statistics Theorem, Low Dimensional Topology and Geometry

6.1 History

We saw in the previous chapter the role of topology in multiparticle systems of elementary particles and providing explanation for possible statistics. It gave the topological origin of bosonic and fermionic statistics in three and higher dimensions. It also provided the possibility of anyonic statistics in two dimensions. But in addition to the statistics of identical particles, there is also the spin of the particles which intervenes in the processes. There is an intriguing relation that all the spin half integral particles obey Fermionic statistics and integral particles have Bosonic statistics. The spin of a particle is related to the rotation group and its representations whereas statistics is for many particle systems. The natural question is why these two should be related and if so how the topology of configuration space and symmetry group of single particle systems together play a combined role. Can this be extended to two dimensions where the new statistics do arise?

The above questions have an interesting history. Pauli was puzzled and provided some answers [1]. In 'Feynman Lectures on Physics', Feynman asked: 'Why is it that particles with half-integral spin are Fermi particles, whose amplitudes add with the minus sign, whereas particles with integral spin are Bose particles whose amplitudes add with the positive sign under exchange?' [2]. Then he goes about apologizing for not providing a simple explanation. Pauli and later Schwinger gave complicated arguments using relativistic quantum field theory and Poincaré symmetry [3].

The basic arguments against these arise from the fact that electrons in atomic orbits even in nonrelativistic regimes need Pauli's exclusion principle to provide stability. There is an interesting paper by Duck and Sudarshan, while providing the background tries to give a reasonably satisfactory answer to this profound question [4].

We provide topological and geometric arguments following the works of Bacry and Broyle [5] and Balachandran et al. [6]. But still, the question remains in the domain of discussions and is waiting for a good explanation.

The arguments essentially look for the classical configuration space where the topological properties are such that the loop corresponding to a rotation by 2π is homotopic to the loop obtained by the exchange of identical particles, i.e.

$$[R(2\pi)\Psi(\vec{r}_1))]\Psi(\vec{r}_2) = \Psi(\vec{r}_2)\Psi(\vec{r}_1) \tag{6.1}$$

6.1.1 Assumptions

1. We assume the elementary particle or even soliton or quasi-particle of quantum field theories come with internal degrees of freedom described by the rotation of the frame of reference.
2. The additional input from relativistic theory we need is, there exists antiparticles and processes where particle and antiparticles can be pair produced from vacuum and annihilated.
3. Multiparticle configurations are provided by the direct product of single particle configurations modulo the permutation group. The last input is needed to implement the identical and indistinguishable nature of particles.

In the previous chapter, we used the topology of such a configuration space to arrive at the possible statistics of particles in three and higher dimensions. Also, it provided the possibility of new statistics in two dimensions.

The configuration of a single particle in three Euclidean dimensions is provided by $\mathbb{R}^3$. In addition, the particle is described by the rotational state as an internal degree of freedom. Hence we attach a frame as a unit vector indicating the direction of the spin pointing. The dynamical evolution of the particle carries such a vector along with it. There could be other degrees of freedom and the spin-statistics theorem assumes the exchange of particles with those which are identical. A natural mathematical description for such a configuration uses the concepts of the fiber bundle. We will avoid using such terms for our purpose.

The time evolution of such a particle without translation with a 2π rotation about an axis (say z-axis) can be picturised as shown in Fig. 6.1.

Consider a process in which two pairs of particle-antiparticles created from vacuum at points $\vec{r}_1, \vec{r}_2$ as indicated in the Figure. Remember the spin components should be same for the particles to continue exchange as indistinguishable and identical nature.

6.1 History

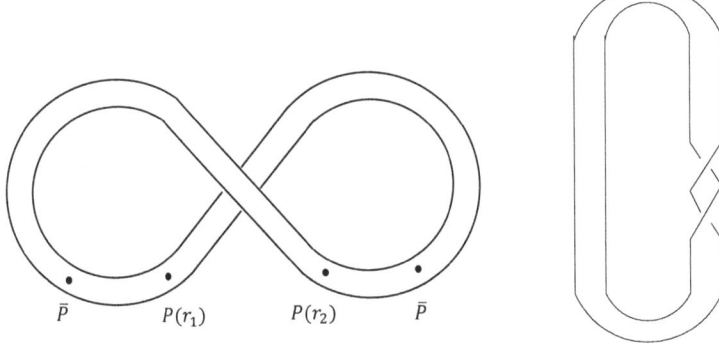

Fig. 6.1 Spin statistics

Now exchange the two identical particles, without any change to the antiparticles. Then complete by annihilating the exchanged particles (**P**) with antiparticles (**P̄**) resulting in vacuum. Hence we have the process

$$vacuum \rightarrow$$
$$\bar{\mathbf{P}}(\vec{r}_1) + \mathbf{P}(\vec{r}_1) \quad \oplus \quad \mathbf{P}(\vec{r}_2) + \bar{\mathbf{P}}(\vec{r}_2)$$
$$exchange$$
$$\bar{\mathbf{P}}(\vec{r}_1) + \mathbf{P}(\vec{r}_2) \quad \oplus \quad \mathbf{P}(\vec{r}_1) + \bar{\mathbf{P}}(\vec{r}_2)$$
$$\rightarrow vacuum$$

We can topologically change this loop continuously to a rotation by 2π of a particle alone created and annihilated along with its antiparticle from vacuum (Fig. 6.1). Since half integral spin particles are described by spinorial representations which pick up negative sign, the exchange of such particles also result in antisymmetry which characterises fermions. This is shown in the Fig. 6.1. Since the complete process hinges on topological equivalence of the two loops and the theorem follows.

Theorem 6.1 $R(2\pi)$ = *Exchange Statistics*

The topological nature along with minimal inputs from relativity (through Poincaré invariance) like the existence of antiparticles and discrete symmetries like CPT and locality, makes this explanation interesting. While the proof is generic in nature, the questions of interactions are unanswered. Also, there are cases where particles and antiparticles are the same.

This can be extended to low dimensional QFTs like two dimensional systems where we have already seen statistics can be anyonic in nature. This we will consider now.

6.2 Anyons and Fractional Spin

In the above section, we saw exchange of two identical particles produces the effect of rotation by 2π of a single particle. But in two space dimensions, we came across that the statistics can interpolate between bosonic and fermionic. This comes through topological features of multiparticle configuration space in two dimensions leading to Artin's braid group [7] and its one-dimensional representations. What about the spin-statistics theorem in that case? Do such anyons have angular momentum described by an angle? Or do they have fractional spin? The answer turns out to be yes as described by Wilzcek, explained in his model [8].

Consider a charge e particle going around a magnetic flux Φ in two dimensions. This can be considered as a model for two anyons at $\vec{r}_1, \vec{r}_2$. We transform to the center of mass coordinate $\vec{R} = \frac{1}{2}(\vec{r}_1 + \vec{r}_2)$ and the relative coordinate $\vec{r} = \vec{r}_1 - \vec{r}_2$. The question is how does rotation act on such a composite system. The flux Φ can be described by the vector potential

$$A_\phi = \frac{\Phi}{2\pi r} \tag{6.2}$$

This is easily seen from the magnetic field due to an infinite solenoid of radius a which is given by the curl of the vector potential, $\vec{A}$:

$$\vec{A} = \frac{\Phi}{2\pi r}\hat{\phi}, \quad (r > a) \tag{6.3}$$

Remember this is a 'closed' but not 'exact' one-form,

$$A = \frac{\Phi}{2\pi r} d\phi$$

given earlier in Chap. 2. Hence the magnetic field dA outside the flux is zero. As described in the previous chapter (particle on a circle) we can eliminate the vector potential by a gauge (singular) transformation. $A \to A - \partial\left(\frac{\phi\Phi}{2\pi}\right)$. Such a gauge transformation produces a phase factor on the charged particle.

$$\Psi'(r,\phi) = \exp\left(\frac{ie\phi\Phi}{2\pi}\right)\Psi(r,\phi) \tag{6.4}$$

6.2 Anyons and Fractional Spin

Hence we have

$$\Psi'(r, \phi + 2\pi) = \left(e^{ie\Phi}\right)\Psi'(r, \phi) \tag{6.5}$$

The allowed angular momenta are $m + \frac{e\Phi}{2\pi}$ thereby giving rise to fractional spin. The above exercise can be extended to two identical anyons in $\mathbb{R}^2$. By moving to the center of mass coordinates and relative coordinates as explained earlier we can see the system is described by fractional spin and fractional statistics. This was as expected from the spin-statistics theorem. Now we shall consider an example of how these turn out in the quantum field theory (QFT) with solitons in $2+1$ dimensions.

6.2.1 $O(3)$ Nonlinear Sigma Model

We had discussed in Chap. 5 the topological properties $O(3)$ nonlinear sigma model and the appearance of soliton solutions. We also remarked that the spin-statistics theorem in this context will be taken up in this chapter. This we will do now. Let us recollect the action functional for this model.

$$S[\vec{\phi}] = \int d^3x \left\{ \frac{1}{2} \partial_\mu \vec{\phi} \cdot \partial^\mu \vec{\phi} + \lambda(\vec{\phi} \cdot \vec{\phi} - 1) \right\}. \tag{6.6}$$

The configuration space consists of all possible maps from $\mathbf{S}^2 \to \mathbf{S}^2$. The solitons are classified by the homotopic class of such maps. The conserved topological current.

$$j^\mu = \frac{1}{12} \epsilon^{\mu\nu\rho} \epsilon^{ijk} \phi_i \partial_\nu \phi_j \partial_\rho \phi_k . \tag{6.7}$$

provides us with a charge

$$Q = \frac{1}{4\pi} \int d^2x \; j_0 \tag{6.8}$$

which gives the soliton number.

Then the question arises what is the spin and statistics of such solitons? Will they also obey spin-statistics theorem?

This question was answered by Wilzcek. From the conserved current j^μ we can write a vector potential A^μ. This is because the equation $\partial_\mu j^\mu$ (with $\mu = 0, 1, 2$) can be compared with $\vec{\nabla} \cdot \vec{A} = 0$.[1] Such a current can be written as

$$j^\mu = \epsilon^{\mu\nu\lambda} \partial_\nu A_\lambda \tag{6.9}$$

---

[1] Note: this implies A can be written as $\vec{A} = \nabla \times \vec{B}$.

We should remember such a 'vector potential' obtained by inverting Eq. (6.9) will be non-local. Given such a vector potential we can write an additional term to the action which is known as the Hopf term

$$H_{hopf} = \frac{\Theta}{2\pi} \int d^3x \, j^\mu A_\mu . \qquad (6.10)$$

6.2.2 Interpretation of Hopf Invariant

In differential topology, the we have already explained that

$$\pi_0[\mathbf{S}^2] = \pi_1[\mathbf{S}^2] = 1, \, \pi_2[\mathbf{S}^2] = \mathcal{Z} = \pi_3[\mathbf{S}^2].$$

The last result can be easily obtained using the short exact sequence. Hence we can construct an integer invariant characterizing the homotopic class of maps from $\mathbf{S}^3 \to \mathbf{S}^2$. We know $\mathbf{S}^3$ is a hypersphere in $\mathbb{C}^2$ with the condition $|z_1|^2 + |z_2|^2 = 1$. We have a map from $\mathbf{S}^3$ to $\mathbf{S}^2$ through

$$\vec{n} = \bar{z}\vec{\sigma}z,$$

where $\vec{\sigma}$ are the Pauli sigma matrices and $z = \begin{pmatrix} z_1 \\ z_2 \end{pmatrix}$. This map provides for every point on $\mathbf{S}^2$ a loop in $\mathbf{S}^3$:

$$z \to e^{i\theta} z$$

will give the same $\vec{n}$. The linking number of two loops for two different points on $\mathbf{S}^2$ is the geometric meaning of the Hopf invariant.

The added term measures homotopic classes $\pi_3(\mathbf{S}^2) = [\mathbf{S}^3, \mathbf{S}^2] = \mathcal{Z}$ mentioned in the Chap. 4. Given a configuration ϕ^i, the Hopf term will turn out to be nonlocal as we mentioned earlier. But the $\mathbf{CP}^1$ formulation of $O(3)$ sigma model (Sect. 5.4.2) where we describe in terms of two complex variables (instead of three real variables) it can be presented as a local term, since we have increased the number of field variables [9].

Since the additional term measures the fundamental group of the configuration space and which is having the same structure $\mathcal{Z}$ like anyonic physics, it provides additional fractional spin and changes the statistics. This will be presented as an exercise.

6.3 Low Dimensional Topology and Geometry

Having described $O(3)$ sigma model as spaces of maps from $\mathbf{S}^2$ to $\mathbf{S}^2$, we can look for extensions beyond that. Let us recollect that $\mathbf{S}^2$ arises through compactification of $\mathbb{R}^2$ due to the physical requirement of finite energy configurations leading to vanishing of the Hamiltonian density at spatial infinity. We can also introduce periodic boundary conditions and look for configurations of maps from $[\mathbf{T}^2, \mathbf{S}^2]$ [10]. This leads us to the classification of compact two dimensional spaces. It is well known they are described by the integer g, the genus of the space which we briefly describe.

6.3.1 Genus g Riemann Spaces

The classification theorem for surfaces says that every compact two-dimensional surfaces is homemorphic to sphere, torus, or connected (glued) sum of tori. The number of tori is the genus of that surface. For sphere $\mathbf{S}^2$ the genus is zero, torus (cycle tube or doughnut)$\mathbf{T}^2$ is the genus one and so on. The sphere can be thought of as a square with all its edges identified as one point. Similarly, torus can be obtained by identifying opposite edges. Genus g surface can be obtained from a 4g-sided polygon by identifying the edges appropriately as shown in the Fig. 6.2. It is clear that the periodic boundary condition of fields on a square will lead to the topology of torus $\mathbf{T}^2$. As mentioned in the Chap. 1, a cube is topologically the same as a sphere. There is an invariant which we immediately get by counting the number of vertices(V), edges(E) and faces(F) of a compact surface of generalized surface S. It is the Euler characteristic $\chi(S)$

$$\chi(S) = 2 - 2g = V - E + F. \tag{6.11}$$

For a smooth manifold of genus g, we can get this invariant as an integral. The surfaces are also described by complex functions and complex geometry.

We will now consider the $O(3)$ model on a torus. The fields $\phi^i(x, y)$ obey $\phi^i(x + L, y) = \phi^i(x, y)$, and $\phi^i(x, y + L) = \phi(x, y)$. Such configuration spaces Γ are characterised by a soliton number. It is easy to show [10] that $\pi_0(\Gamma) = \mathcal{Z}$. This is the same as the previous case of maps of $\mathbf{S}^2 \to \mathbf{S}^2$. The same conservation law holds good here also. Next, what is the fundamental group of the configuration space, viz.,

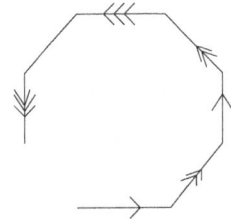

Fig. 6.2 Genus g surface

$\pi_1(\Gamma)$. To obtain this we focus on the soliton number zero sector. In this sector, we can have a constant map. Hence we have loops starting from $\phi = (0, 0, 1)$ and ending with the same vector. We can define two topological integers m_x, m_y such that

$$m_x = \frac{1}{4\pi} \int dy\, dt\, \epsilon^{abc}\, \phi^a \partial_y \phi^b \partial_t \phi^c \tag{6.12}$$

and m_y being similarly defined. They can be given by the generators X, Y. This gives homomorphism between $\pi_1(\Gamma)$ and $\mathcal{Z} \oplus \mathcal{Z}$. The question is what is $X Y X^{-1} Y^{-1}$ or the commutator $[X, Y]$. It can be shown to be $[X, Y] = Z^2$ where the third generator Z is constructed using the Hopf invariant of the previous section. Hence we have three generators X, Y, Z in the zero soliton sector obeying the relations

$$XZ = ZX, \quad YZ = ZY, \quad XY - YX = Z^2 \tag{6.13}$$

It is interesting such a structure emerges for the fundamental group in the zero soliton sector. Working out the fundamental group in higher soliton sectors is more complicated.

The interesting result is the possibility of nonabelian fundamental group in a field theory of solitons and thereby non-abelian statistics [10]. Such a possibility of non-abelian statistics have been seen in other examples too recently [11].

Arbitrary Genus Surfaces

The above results can be generalised to arbitrary genus g surfaces. For example, genus g surface $\mathcal{M}_g$ can be obtained from $4g$-sided polygons with proper identifications. The solitons for such a model will be classified by $\pi_0(\Gamma)$. But now Γ is given by maps from $\mathcal{M}_g \to \mathbf{S}^2$. It turns out that $[\mathcal{M}_g, \mathbf{S}^2]$ is also $\mathcal{Z}$. But the $\pi_1(\mathcal{M}_g \to \mathbf{S}^2)$ in the zero soliton sector is obtained by $2g + 1$ generators Z, X_a, Y_a, $a = 1, 2, \ldots g$. It is easy to visualize we will need $2g$ generators exactly along the same lines of argument for the torus. We need exactly one more generator Z which will be the commutator of any pair of generators X_a, Y_a. It is easy to see that this generator will commute with all the $2g$ generators. They obey the group multiplication:

$$X_a X_b = X_b X_a \quad Y_a Y_b = Y_b Y_a$$
$$X_a Z = Z X_a \quad Y_a Z = Z Y_a$$
$$X_a Y_b = Z^2 Y_a X_b, \text{(if } a = b), \quad X_a Y_b = Y_b X_a \text{(if } a \neq b)$$

The fundamental groups of the configuration spaces for the $O(3)$ nonlinear sigma model on the compact genus g surfaces $\mathcal{M}_g$ have been given for any soliton number N. So are the braid group for N spinless particles on these manifolds. The representations of these groups govern the possible statistics of solitons and particles.

It can be shown when spin and creation/annihilation processes are introduced, the fundamental groups for the particles are the same as the corresponding σ-model groups. These fundamental groups incorporate the spin-statistics connection and are of greater physical relevance than the standard braid groups [12].

6.4 Three-Dimensional Geometry and Topology

Having seen some of the quantum physics applications of two dimensional geometry and topology, we will consider three-dimensional geometries and topologies. We already came across the braid group from two-dimensional geometries. It is intuitively clear if we close the braids in two dimensions we will get knots in three dimensions. Hence we can expect close connections between three-dimensional topologies and two-dimensional theories. Interest in topology and geometry of three-manifolds came up with the works of V F R Jones who used statistical Mechanics models [13] to develop his polynomial invariants in three-manifolds as well as E Witten [14] who obtained the polynomial invariants from Chern Simons gauge theories for non-abelian groups $SU(2)$, $SU(N)$. Before going to the details we will explain the classification issues of three-dimensional compact and noncompact manifolds.

6.4.1 Three Manifolds

All three-manifolds are spaces which are locally described by three coordinates. These can be triangulated as simplicial complex as described in Chap. 2. There is an important conjecture due to Poincare made more than 100 years ago which says that every compact simply connected 3-manifold is homeomorphic to $\mathbf{S}^3$. This was proven only recently by Gregori Perelman [15]. There is a geometrization conjecture due to Thurston which says that every closed three-manifold can be decomposed in a canonical way to one of the eight geometric structures. Understanding the geometry and topology of three-manifolds is important from the point of quantizing gravity and other issues. First, we will collect the various types of 3-manifolds.

1. $\mathbf{S}^3$ is the simplest example other than $\mathbb{R}^3$. It is a surface in $\mathbb{R}^4$ given by $\sum_1^4 x_i^2 = 1$ where x_i's are real. It can also be given as $\sum_1^2 |z_i|^2 = 1$ in terms of two complex numbers and has constant curvature. It is also the group manifold of $SU(2)$.
2. If Γ is a group that acts freely and isometrically (i.e. preserving the distance between points) then we get the coset spaces $\frac{\mathbf{S}^3}{\Gamma}$ which is the Riemannian manifold with constant curvature. For example, if p, q are coprimes and

$$z_1' = e^{\frac{2\pi i}{p}} z_1, \quad z_2' = e^{\frac{-2q\pi i}{p}} z_2$$

then we obtain the Lens space $L(p,q)$. Using our earlier results we can get $\pi_1(L(p,q)) = \mathcal{Z}_p$.
3. We also have hyperbolic 3-manifolds $\mathbb{H}^3$ defined by $\{x, y, z\}$, $z > 0$, the upper half space with the metric $ds^2 = \frac{1}{z}(dx^2 + dy^2 + dz^2)$.
4. We get $\frac{\mathbb{H}}{\Gamma}$ as the hyperbolic 3-manifold if Γ acts freely and isometrically.
5. We can have a knot K in $\mathbf{S}^3$. Consider the exterior of the knot $(\mathbf{S}^3/K)$. We can construct many three manifolds using Dehn surgery for any knot K or any link L.
6. One can construct, from two-dimensional genus g surface M_g, three-manifolds $M_g \otimes \mathbf{S}^1$.

Having given a glimpse of all three-manifolds, we can move on to studying the topological and geometric properties using gauge theories. There is a well-known program using Chern-Simons gauge theory which will be our focus.

6.5 Chern-Simons Gauge Theory on $\mathbf{S}^3$

Three-dimensional Chern-Simons (CS) gauge theory is studied from the point of view of three-manifold invariants as well as the theory of knots and links. It was studied in relation to fractional statistics as well as linking numbers. In Chap. 7, we will elaborate on this. With the non-abelian gauge group , more topological invariants of three manifolds and knots/links are brought out. This was initiated by E. Witten [14]. We will formulate the questions here and follow it up in the next three chapters.

The metric-independent Chern-Simons action for U(1) gauge theory is:

$$S = \frac{1}{2\pi} \int A \wedge dA \qquad (6.14)$$

It is metric independent since the volume form $\sqrt{g}\, d^3x$ automatically provides the scalar density. We will explain in the next chapter how this leads to topological invariants like linking numbers. The CS action for non-abelian gauge theory on a three-manifold

$$S = \frac{k}{4\pi} \int_{M^3} tr\left(A \wedge dA + \frac{2}{3} A \wedge A \wedge A\right). \qquad (6.15)$$

Here A stands for $SU(2)$ Lie-algebra valued one form. This means

$$A = A_\mu^a\, T^a\, dx^\mu, \qquad (6.16)$$

6.5 Chern-Simons Gauge Theory on S^3

where T^a stand for $SU(2)$ generators in the adjoint representation. The 'tr' stands for trace over the generators and is normalised as

$$2 \, tr \, T^a \, T^b = \delta^{ab}. \tag{6.17}$$

We can have any compact semisimple Lie algebra, but for clarity, we consider $SU(2)$. First, this action is topologically invariant as it is independent of the metric. Secondly using the fact that $\pi_3(SU(2))$ (for that matter for any compact semisimple Lie group) is $\mathcal{Z}$, we will arrive at the integrality of k. This happens because the action is invariant under small gauge transformations containing identity, but not under large gauge transformations where it changes by $2\pi \, k$. Under gauge transformations

$$A \to g^{-1} \, A \, g + g^{-1} \, d \, g \tag{6.18}$$

where $g \in SU(2)$ we get

$$S \to S + k \int tr \, (g^{-1} \, d \, g)^3 = S + 2\pi k \tag{6.19}$$

Following Dirac's arguments for monopole, we need to make sure that the action S changes not more than $S + 2\pi \, \mathcal{Z}$. This provides the argument for $k = \mathcal{Z}$.

We also need observables which are topologically and gauge invariant quantities. The partition function, as defined in an quantum field theory,

$$\mathcal{Z} = \int \mathcal{D}A \, e^{iS}, \tag{6.20}$$

is a topological invariant which is also a gauge invariant quantity. Here the important technical question is what is the measure $\mathcal{D}A$. For this, we can fix a convenient gauge $A_0^a = 0$. Then A_x^a and A_y^a will become canonical conjugates by the usual procedure from the Lagrangian $\frac{\delta \mathcal{L}}{\delta A_x}$. The canonical commutation relations will be:

$$[A_i^a(x), A_j^b(y)] = \frac{2\pi}{k} \delta^{ab} \epsilon_{ij} \delta^3(x - y) \tag{6.21}$$

Since the equations of motion will turn out to be

$$F = dA + A \wedge A = 0 \tag{6.22}$$

the solution space will be what are known as flat connections (i.e.: the field strength is zero). They are given by $A = g^{-1} \, dg$. The quantization is to be carried out in such a phase space. The gauge invariant observables are Wilson loop operators:

$$W_R[C] = tr_R \, P \, \exp \oint_C A \tag{6.23}$$

The [C] stands for a loop (which could be an oriented knot). The symbol P stands for path ordering. This is needed since the gauge field A does not commute for different positions. Since the theory is topologically invariant the expectation values of the operator could depend only on the topological characterization of the knot as well as the representation R that lives on the loop.

We can generalise the above observable to links $L(C_1, C_2, ..)$:

$$W_{R_1,R_2,...}[L] = \prod_i W_{R_i}[C_i] \qquad (6.24)$$

We can also place the same representation in all the components $[C_i]$ where the observable is $W_{R,R,...}[L]$. The functional average of the observable, as in any QFT, is:

$$V_R[L] = \langle W_R[L] \rangle \equiv \frac{\int \mathcal{D}A \prod_i W_R[C] e^{iS}}{\int \mathcal{D}A \, e^{iS}} \qquad (6.25)$$

Since the measure, integrand in the functional integral are independent of the metric, it is obvious that it can depend only on the knot/link characterization as well as the representations R_i of the gauge group used. Having set up the question in a topological setting, we will explain in Chap. 7 the mathematical theory of knots, links and three-manifolds in CS gauge theoretic formulation. Then we will explore how the above expectation values can be computed. Interestingly the above expressions can be analytically computed for the gauge group $SU(2)$ with any representation R living on the components of the links. We will provide examples of such computations.

Exercises

6.1 A typical ansatz for rotationally symmetric configuration would be $\vec{\phi} = (0, 0, -1)$ at origin and $(0, 0, 1)$ as $r \to \infty$. Write down such a configuration and verify the soliton number given by Eq. (6.8) is one.

6.2 Show that in $O(3)$ nonlinear sigma model with Hopf term, the soliton number one solution has fractional angular momentum given by Θ the coefficient of the Hopf term.

6.3 Write down $O(3)$ sigma model action in terms of **CP**^{1} variables and show it is completely local.

6.4 Show $\pi_0(\Gamma) = \mathcal{Z}$ where Γ is the space of maps from $\mathbf{T}^2 \to \mathbf{S}^2$.

6.5 For Γ the same space of maps as above from $\mathbf{T}^2 \to \mathbf{S}^2$, show $\pi_1(\Gamma)$ is given by three generators X, Y, Z obeying Eq. (6.13).

6.6 Consider Yang-Mills gauge theory with the group $SU(N)$. Local gauge transformations are those which go to identity at asymptotic infinity. Global gauge transformations are those that are all gauge transformations modulo local transformations. Using the above definitions obtain the $\pi_0(\mathcal{Q})$, $\pi_1(\mathcal{Q})$ where $\mathcal{Q}$ is the configuration space of observables. Explain how it is related to the strong CP problem in QCD.

References

1. W. Pauli, Phys. Rev. **58**, 716 (1940)
2. R.P. Feynman, R.B. Leighton, M. Sands, *The Feynman Lectures on Physics*, vol. 3, Chap. 4 (Addison-Wesley publisher, 1965)
3. J. Schwinger, Phys. Rev. **82**, 914 (1951)
4. I. Duck, E.C. George Sudarshan, Am. J. Phys. **66**, 284 (1998)
5. H. Bacry, Am. J. Phys. **63**(4), 297 (1995)
6. A.P. Balachandran, A. Daughton, Z.-C. Gu, R.D. Sorkin, G. Marmo, A.M. Srivastava, Int. J. Mod. Phys. A **8**, 2993 (1993)
7. E. Artin, Ann. Math. **48**, 101 (1947)
8. F. Wilczek, Phys. Rev. Lett. **49**, 957 (1982)
9. F. Wilczek, A. Zee, Phys. Rev. Lett. **51**, 2250 (1983)
10. T.R. Govindarajan, R. Shankar, Mod. Phys. Lett. **A4**, 1457 (1989)
11. M. Banerjee, M. Heiblum, V. Umansky, D.E. Feldman, Y. Oreg, A. Stern, Nature **559**, 205 (2018)
12. A.P. Balachandran, T. Einarsson, T.R. Govindarajan, R. Ramachandran, Mod. Phys. Lett. A **6**, 2801 (1991)
13. V.F. Jones, Bull. Am. Math. Soc. (N.S.) **12**, 103 (1985)
14. E. Witten, Commun. Math. Phys. **121**, 351 (1989)
15. G. Perelman, (arXiv paper math/0211159)

Part III

Braid Group, Knots, Three Manifolds

We introduced braids and their group structure in Chap. 5. We briefly recollect here so that we can study how knots in three-dimensional space $\mathbb{R}^3$ arise from these braids. We then provide a brief historical introduction to the theory of knots from a mathematical and physical point of view. Then we will present how this approach helps us both in terms of physical applications and mathematical understanding of questions for the classification of knots and three-dimensional manifolds. This interaction between Physics and Mathematics turned out to be useful from both perspectives. But the theory of knots has deeper necessity in terms of approaches to quantising gravity [1] itself. Now we move on to the theory of braids.

7.1 Braids, Braid Group

We saw Artin's braid group $\mathcal{B}_N$ (introduced in Chap. 5) [2] is generated by $N-1$ generators b_i, $i = 1, 2, \ldots N - 1$. The generator b_i correspond to exchanging braids $i \leftrightarrow (i+1)$ with i-th strand going over $i+1$ strand. These generators obey the relations:

$$b_i\, b_j = b_j\, b_i \quad |i - j| \geq 2,$$
$$b_i\, b_{i+1}\, b_i = b_{i+1}\, b_i\, b_{i+1}.$$

There are several representations for the braid group, but we specifically used $b_j = e^{i\theta}$ earlier to explain the exotic fractional statistics [3] of particles in two-dimensional quantum systems. These one-dimensional representations were considered, since in quantum theory, the multiparticle wave function $\Psi(r_1, r_2, \cdots, r_N)$ is a map $\mathbb{R}^N \to \mathbb{C}$. Such phase factor proved to be useful in the integer and fractional quantum hall effect involving fractional statistics of the exotic particles in the two-dimensional space. We also want to recall that the braid group structure closely resembles the Yang-Baxter (YB) structure in integrable statistical mechanics problems. Yang-Baxter equations have different forms, but the one directly linked to the braid group is described by a matrix $\hat{R}$ which acts on the Hilbert space $\mathcal{H} \otimes \mathcal{H}$. Then the consistency conditions on $\hat{R}$ matrix[1] for integrability turns out to be on the tensor product $\mathcal{H} \otimes \mathcal{H} \otimes \mathcal{H}$ as:

$$(\hat{R} \otimes 1)(1 \otimes \hat{R})(\hat{R} \otimes 1) = (1 \otimes \hat{R})(\hat{R} \otimes 1)(1 \otimes \hat{R}) . \tag{7.1}$$

One can immediately see the similarity between YB equations for integrability and braid group. This gives us a method of obtaining representation of the braid group from representation for $\hat{R}$ in an integrable statistical mechanics model. There is also one parameter-dependent form of Yang-Baxter condition for integrable systems which also provide braid group representations. We will remark on this briefly.

Lastly, in two-dimensional conformal field theories too, there appear what is known as monodromy matrices when a field is transported along a loop around another [4]. These too are related to the braid group generators and will provide a representation of braid groups. We will be extensively using these to obtain several interesting relations between braid group representations and the knots in three-manifolds [5].

7.1.1 Knots

Knots are loops in $\mathbb{R}^3$ or in any other three-manifolds. These are the embedding of $\mathbf{S}^1$ inside the three-manifolds. Can we go from one knot to another without cutting and joining the loop is a topological question. Such transformations define the topological equivalence of knots.

While a knot exists in three dimensions, we can describe it as a planar projection with either over-crossing or under-crossing. In general, there is no direction associated with the knot but we can provide an arrow along the knot also. These are known as oriented knots. Similarly, links can be drawn as many component knots. Knots have been useful historically, and people have used several types in day-to-day life. But in Physics, it was proposed first by Kelvin [6] to indicate the chemical elements in the periodic table as vortex lines knotted in the now discarded notion of ether medium. Tait obtained various topologically inequivalent knots after removing those that could be obtained from one to another by pulling and/or pushing without cutting and pasting.

Mathematically two knots K_1, K_2 are equivalent if there exist an orientation preserving homeomorphism $\mathcal{O}$ such that $\mathcal{O}(K_1) = K_2$ There is a set of three moves which can be done to deduce the topological equivalence amongst knots. That is.,

---

[1] R matrix encodes Boltzmann weights in lattice models and satisfies Yang-Baxter (YB) equation for solvability.

7.1 Braids, Braid Group

two knots are equivalent if by a suitable combination of these moves, one knot can be deformed to the other knot. These are known as Reidemeister moves:

1. twist or untwist a part of strand in any direction.
2. Move a strand over another completely.
3. Move a strand completely over an under/over crossing made by two other strands.

They are explicitly drawn below.

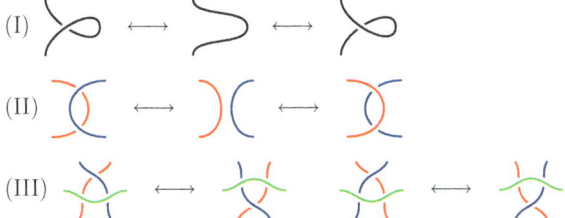

7.1.2 Some Examples

We draw a few examples of knots and links (see Figs. 7.1 and 7.2).

Then we present the knot polynomials/knot invariants associated to the knots and links in the following subsection.

7.1.3 Knot Invariants

It is a quantity or polynomial we associate to distinguish different knots. If the associated quantities are different for two knots, then we can say that those knots

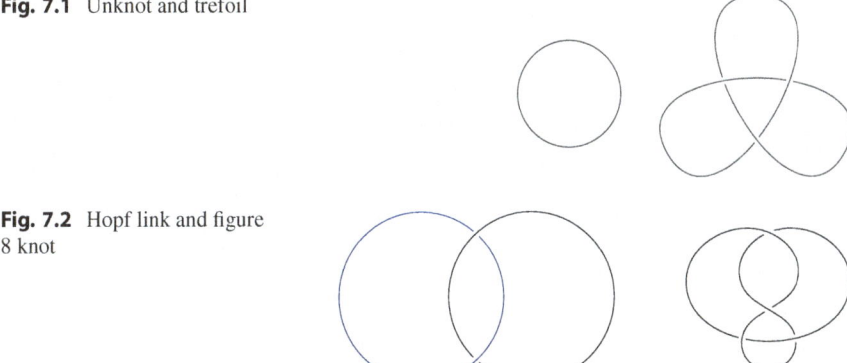

Fig. 7.1 Unknot and trefoil

Fig. 7.2 Hopf link and figure 8 knot

are not equivalent. It may happen that two different knots (not related by the Reidemeister moves) have the same invariant.

Alexander and Conway associated a polynomial in the variable q with integer coefficients with an oriented knot. The rules for obtaining the polynomials are stated on the oriented knot diagram on a plane with appropriate over and under-crossing. Focus on a crossing in an oriented knot diagram with an over-crossing (indicated as subscript + and knot K_+). The knot invariant of the diagram $V_+(K_+)$ is related through a skein relation of invariants of two diagrams K_-, K_0 obtained by replacing the over-crossing by (1) under-crossing $V_-(K_-)$ and (2) by no crossing $V_0(K_0)$:

$$V_+(K_+) - V_-(K_-) = (q^{1/2} - q^{-1/2}) V_0(K_0), \qquad (7.2)$$

with the unknot $\bigcirc$ polynomial taken as $V(\bigcirc) = 1$. Such a skein or recursion relation will result in a polynomial for K_+ provided we know the polynomial invariant for K_- and K_0. The following example shows explicitly the polynomials for trefoil (T) and Hopf(H) using the above skein relation.

$$V(T) = V(\bigcirc) + (q^{1/2} - q^{-1/2})V(H)$$
$$V(H) = V(\bigcirc \cdot \bigcirc) + (q^{1/2} - q^{-1/2})V(\bigcirc).$$

We leave the readers to check that the polynomial is zero for any disjoint knots. Hence, the polynomial for two disjoint unknots will be $V((\bigcirc \cdot \bigcirc)) = 0$. Substituting in the above two equations, the Alexander polynomial for Hopf link and trefoil are

$$V(H) = (q^{1/2} - q^{-1/2}),$$
$$V(T) = q - 1 + q^{-1}.$$

We get different polynomials for trefoil, Hopf link and unknot confirming that they are not equivalent. Though visually this is a simple statement, it is interesting we get them through the algebraic relations. But, the trefoil knot T and its mirror image T^* as drawn in Fig. 7.3 are inequivalent. That is., there are no Reidemeister moves connecting trefoil T to its mirror image T^*. However, the Alexander polynomial turns out to be the same. These two knots are sometimes referred to as left-handed and right-handed trefoil. This problem of distinguishing knots from their mirror images remained an outstanding question for nearly sixty years. This was resolved by V F R Jones through a polynomial(named after him) with a modified skein

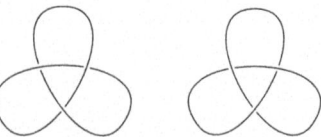

Fig. 7.3 Right and left handed trefoil

relation. This new development came through the study of integrable statistical mechanics model. The new skein relation is:

$$q^{-1}J_+(K_+) - q J_-(K_-) = (q^{-1/2} - q^{1/2})J_0(K_0) , \qquad (7.3)$$

where $J(K)$ stands for Jones polynomial for a oriented knot K. In fact, $J(T) \neq J(T^*)$ indicating that the Jones polynomials are better than Alexander polynomial. To match with the physics notations, we will change the unknot normalization. This can be achieved by requiring the knot invariant of two disconnected knots is a product of knot polynomials.

$$J(K_1 \# K_2) = J(K_1)J(K_2) . \qquad (7.4)$$

This immediately leads to (considering two disconnected unknots) the polynomial for unknot as:

$$J(\bigcirc) = q^{1/2} + q^{-1/2} . \qquad (7.5)$$

The significance of the above polynomial from the $SU(2)$ Chern-Simons theory will be pointed out later. Using the skein relation we can obtain knot invariants of several knots. For example:

1. Trefoil: $J(T) = (q + q^3 - q^4)J(\bigcirc)$.
2. Hopf link: $J(H) = (q^{5/2} + q^{1/2})J(\bigcirc)$.
3. Figure 8 (knot 4_1): $J(4_1) = (q^2 - q + 1 - q^{-1} + q^{-2})J(\bigcirc)$.

Note that the Alexander Conway polynomial for trefoil is symmetric under the exchange $q \leftrightarrow q^{-1}$ as derived from Eq. (7.2). But, the Jones polynomial is not. This is related to the fact that the Alexander-Conway polynomial for trefoil does not distinguish between trefoil T and anti-trefoil T^*. But Jones polynomial does. The knot inequivalence of T, T^* is established from the Jones polynomial

$$J(T) \xrightarrow{q \text{ to } q^{-1}} J(T^*) \neq J(T) .$$

Following the work of Jones, two more important knot polynomials were given. One is known as Kauffman polynomial [7] and the other as HOMFLY-PT polynomial. HOMFLY-PT polynomial was given independently by Hoste, Ocneanu, Millett, Freyd, Lickorish and Yetter [8] and separately by Przytycki and Traczyk [9].

The Kauffman polynomial is $F(L) = a^{-\omega(L)}K(L)$ where $K(L)$ obeys:

$$K(L_+) + K(L_-) = z(K(L_0) + K(\bar{L}_0)) . \qquad (7.6)$$

In the above $\omega(L)$ is the writhe (which we define later in this chapter). L_+ denotes the link with a chosen crossing as over-crossing whereas L_- corresponds to the link with that crossing made under-crossing. The above skein relation is applicable for unoriented diagrams. Hence, there are two possible no-crossing diagrams L_0 and its dual no-crossing $\bar{L}_0$. Thus the Kauffman polynomial is a polynomial in two variables a, z for unoriented knots and links.

HOMFLY-PT polynomial is another two variable polynomial, in variables w, q, given by the skein relation amongst oriented knots/links:

$$\omega^{-1/2} q^{-1/2} P_+(K_+) - \omega^{1/2} q^{1/2} P_-(K_-) = (q^{-1/2} - q^{1/2}) P_0(K_0) \,. \tag{7.7}$$

The above skein relation reduces to the Jones skein relation for $\omega = q$ and Alexander-Conway relation for $\omega = q^{-1}$.

We rewrite the above by defining $\omega = q^{N-1}$ as:

$$q^{-N/2} H_+ - q^{N/2} H_- = (q^{-1/2} - q^{1/2}) H_0 \,. \tag{7.8}$$

We will see the above rewriting helps us to compare with $SU(N)$ Chern-Simons theory.

It is easy to see that the HOMFLY-PT polynomial for unknot is:

$$H(\bigcirc) = \frac{\omega^{-1/2} q^{-1/2} - \omega^{1/2} q^{1/2}}{q^{-1/2} - q^{1/2}} \,. \tag{7.9}$$

In terms of q and N the above polynomial is

$$H(\bigcirc) = \frac{q^{N/2} - q^{-N/2}}{q^{1/2} - q^{-1/2}} \,. \tag{7.10}$$

This unknot polynomial reduces to the Jones polynomial for unknot when we take $N = 2$.

Having given a brief introduction to the theory of knots from mathematical developments, we will focus on how we can obtain the same using Chern-Simons theory on $\mathbf{S}^3$ based on simple gauge groups like $U(1)$ and $SU(2)$.

7.2 Chern Simons Theory and Knot/Link Invariants

In the previous Chap. 6, we introduced Chern Simons (CS) gauge theory on three-manifolds. We will develop further here explaining how the knot/link invariants arise in this theory. First, we will consider the simplest case of the $U(1)$ gauge group and explore the linking invariants that can be described through this model. We can also obtain a self-linking invariant of a knot from this theory.

7.2.1 $U(1)$ Chern-Simons Theory and Linking Invariants

$U(1)$ CS theory is an abelian theory which is defined by the following action:

$$S = \frac{k}{4\pi} \int_{S^3} A \wedge dA = \frac{k}{4\pi} \int \epsilon^{\mu\nu\rho} A_\mu \partial_\nu A_\rho \, d^3x \,. \tag{7.11}$$

For clarity, we have taken the three-manifold to be $\mathbf{S}^3$. We can provide an invariant associated with links, say made of two knots C_1, C_2. The linking number is an invariant associated with links up to isotopy. For this, we associate whenever there is a crossing of two oriented strands $+$ and $-$ sign for overcrossing and undercrossing respectively. Then, half the difference between the number of $+$ crossings N_+ and the number of $-$ crossings N_-: $1/2(N_+ - N_-)$ between the two components C_1, C_2 is an invariant known as linking number $Lk(C_1, C_2)$. For r number of component knots making the link diagram D, the linking number of the link D is

$$Lk(D) = \sum_{1 \le i < j \le r} Lk(C_i, C_j) \,. \tag{7.12}$$

$Lk(C_i, C_j)$ measures the number of times the curve C_i goes through the curve C_j.

Examples
1. Two unknots $Lk(\bigcirc, \bigcirc) = 0$.
2. Hopf link $Lk(H) = 1$ (both the knots are clockwise oriented).
3. Borromean rings involve interlocking of three curves (See Fig. 7.4). The linking number will be zero.

Writhe We can introduce another quantity for a knot or link known as writhe. This is simply the sum of signs of crossings. A trefoil has writhe 3, Hopf link has writhe 2 and Borromean ring writhe 0. Note that the writhe changes under the first Reidemeister move of twisting many times a strand. Hence writhe is not an ambient isotopy invariant.

Fig. 7.4 Borromean rings

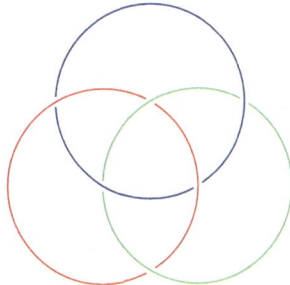

We will now see how these invariants, linking number and writhe, are obtained from the $U(1)$ CS theory. The $U(1)$ CS partition function involving the action in Eq. (7.11) is

$$Z[\mathbf{S}^3] = \int \mathcal{D}A \, e^{iS}, \qquad (7.13)$$

where $\mathcal{D}A$ stands for the measure in the space of all gauge potentials modulo the gauge transformation. We can write the observable, which is gauge invariant and topological invariant, as Wilson loops are given by knots embedded in the three-manifold $\mathbf{S}^3$. For example given a link made up of two knots C_1, C_2 we have,

$$W(C_1, C_2) = \exp\left\{i \oint_{C_1} A_\mu \, dx^\mu \oint_{C_2} A_\nu \, dy^\nu \right\}. \qquad (7.14)$$

The expectation value of the observable is defined as:

$$<W(C_1, C_2)> = \frac{\int \mathcal{D}A \, e^{iS} \, W(C_1, C_2)}{Z[\mathbf{S}^3]}. \qquad (7.15)$$

The above expression is not very difficult to compute, since it involves only a quadratic or Gaussian integral. We can solve the classical equations through a source term $\int j_\mu A^\mu$ where the current $j(x)$ is due to a point particle. The equations are:

$$\vec{\nabla} \times \vec{A} = \frac{2\pi}{k} \vec{j}. \qquad (7.16)$$

By taking curl on both sides we get:

$$\nabla^2 \vec{A} = -\frac{2\pi}{k} \vec{j}. \qquad (7.17)$$

Using the current $j^\mu(\vec{y}) = \int dx^\mu \, \delta^3(x-y)$ we get the expectation value as:

$$<W(C_1, C_2)> = \frac{4\pi i}{k} Lk(C_1, C_2), \qquad (7.18)$$

where $Lk(C_1, C_2)$ is:

$$Lk(C_1, C_2) = \frac{1}{4\pi} \oint_{C_1} dx^\mu \oint_{C_2} dy^\nu \epsilon_{\mu\nu\lambda} \frac{(x-y)^\lambda}{|x-y|^3}. \qquad (7.19)$$

Thus we get the linking invariant of the two knots. This can easily be generalised to any link.

Self-linking Invariant

We can provide a self-linking number for knots also. That requires careful analysis. First, the two knots should overlap. That is $C_1 = C_2$. For this, we consider $C_2 = C_1 + \epsilon\, y$. That is C_2 is close to C_1 and obtained by displacement y also known as its frame. Then the expectation value, in the limit of $\epsilon \to 0$ is given by $\frac{4\pi i}{k} SL(C_1)$ where $SL(C)$ is the self-linking number of the curve which depends on the framing of the loop C. Since abelian $U(1)$ theory is a free theory and quadratic in field variables, we can explicitly obtain expectation value for any observable. Linking and self-linking numbers provide the complete description of this theory. To go beyond these linking invariants, we have to study an interacting CS theory. We will now discuss such a theory.

7.2.2 $SU(2)$ Chern Simons Theory on S^3

The CS theory based on a nonabelian gauge group will have cubic interaction terms in the action. So, the computation of expectation values of observables becomes difficult perturbatively. But this provides an invariant at every order in the CS perturbative expansion and the sum through all orders gives us a polynomial invariant. It is indeed possible to follow nonperturbative methods directly and obtain polynomial invariants. We will describe such a procedure now.

Now we will develop further the methods for exactly computing topological invariants which are expectation values of observables.

7.2.3 Canonical Quantisation

Gauge theory quantisation requires proper identification of phase space variables after imposing constraints on the Hilbert space. Only those operators corresponding to these observables act on the Hilbert space. First, we realise that the functional integral given by the partition function $Z[\mathbf{S}^3]$ itself must be a topological quantity. This is because the action does not contain a metric and is invariant under transformations of the metric. These are known as Schwarz type topological theory [10]. Next, we split the manifold M into two spaces M_1, M_2 and glue them to get back the manifold. These spaces will have two-manifold boundaries like for example $\Sigma = \partial M = \mathbf{S}^2$. Hence the functional integral will turn out to be a state-functional on the boundary.

$$|\Psi(\partial M)\rangle = Z(M, \partial M) = \int \mathcal{D}A\, e^{iS} \qquad (7.20)$$

Near the boundary the three-manifold has the structure $\Sigma \times \mathbb{R}$. Here $\mathbb{R}$ plays the role of time. For convenience we choose the gauge $A_t = 0$. This reduces the action to

$$S = \frac{k}{8\pi} \int dt \int_\Sigma tr A \wedge \dot{A} \, . \tag{7.21}$$

This makes the two components of the gauge potential as canonically conjugate.

$$\{A_i^a(x), A_j^b(y)\} = \frac{4\pi}{k} \epsilon_{ij} \delta^{ab} \delta^2(x-y) \, . \tag{7.22}$$

We also have the equations of motion (Gauss constraint):

$$\epsilon^{ij} F_{ij}^a = 0 \, . \tag{7.23}$$

Now we have an option of following Dirac's theory for constrained systems. We can quantise by finding the representation of the canonical commutation relations on a Hilbert space and then choose those states which are restricted by the Gauss constraint equation. Or we can find the independent observables by imposing the constraint and then quantise the algebra obeyed by them. Now the constraint is that the field strength is zero. This implies that we have the space of flat connections. The solution for $F = dA + A \wedge A = 0$ is $A = g^{-1}dg$.

Hence the space of solutions correspond to the space of flat connections (one form). But we have to identify those connections which are related by gauge transformations. This fixes the phase space to be space of flat connections modulo gauge transformations. There are two terms that can be written using the flat connections on the two dimensional manifolds ∂M. The flat connection is one form $A = g^{-1}dg$. We can write down gauge invariant term

$$Tr \, g^{-1}dg g^{-1}dg = Tr(g^{-1}\partial_\mu g \, g^{-1}\partial^\mu g) \, , \tag{7.24}$$

which is the kinetic term of a two-dimensional sigma model. There is also another topological term known as a Wess-Zumino term given by:

$$Tr \left(\epsilon^{\mu\nu\lambda} g^{-1}\partial_\mu g g^{-1}\partial_\nu g \cdot g^{-1}\partial_\lambda g \right) \, . \tag{7.25}$$

Combining both we can write an action function on ∂M

$$S = k \left(\frac{1}{16\pi} \int Tr(\partial_\mu g \, \partial^\mu g^{-1}) + \frac{1}{24\pi} \int Tr(g^{-1}dg)^3 \right) \, . \tag{7.26}$$

The second integral is over a three-manifold whose boundary is given by ∂M. The Hilbert space of flat connections, which are one forms valued in the Lie algebra of the gauge group G on Riemann space ∂M, is also related to that obtained for a conformal field theory whose action is given above. The representation space can

7.2 Chern Simons Theory and Knot/Link Invariants

be constructed by looking at the current algebra symmetry obtained from the above action. That symmetry is known as Kac-Moody algebra and the integer k we have in the action turns out to be the level of the algebra. We will briefly describe the algebra.

7.2.4 Brief Introduction to the $SU(2)_k$ Kac-Moody Algebra

The affine $SU(2)_k$ Kac-Moody algebra [11] also known as the current algebra is described by the currents J_m^a, $m \in \mathcal{Z}$, $a = 1, 2, 3$:

$$[J_m^a, J_n^b] = \epsilon^{abc} J_{m+n}^c + kn\delta^{ab}\delta_{m+n,0}, \tag{7.27}$$

where we have explicitly used $SU(2)$ structure constants. (Otherwise replace $\epsilon^{abc} \to C^{abc}$). This algebra is associated with holomorphic currents $\partial_z g g^{-1}$. The antiholomorphic currents provide another chiral algebra $\bar{J}_m^a$. The full symmetry of the WZW model is given by two copies of the affine Lie algebra. The spectrum of the WZW at level k is given by

$$\Phi(k) = \oplus \mathcal{R}_j \otimes \bar{\mathcal{R}}_j. \tag{7.28}$$

In the above, $j = 0, \frac{1}{2}, 1, \cdots, \frac{k}{2}$. These are known as the primary fields of CFT. We obtain the Virasoro algebra by what is known as Sugawara construction [12] where generators L_n are:

$$L_n = \frac{1}{2(k+2)} \sum_a \sum_m J_{n-m}^a J_m^a. \tag{7.29}$$

We can check that these generate Virasoro algebra given the current algebra Eq. (7.27). The conformal dimension of the primary filed Φ_j[2] is given by

$$h_j = \frac{j(j+1)}{k+2}. \tag{7.30}$$

The fusion rules when we combine two primary fields $[\Phi_j]$ are

$$[\Phi_{j_1}] \cdot [\Phi_{j_2}] = \oplus N_{j_1, j_2, J} [\Phi_J], \tag{7.31}$$

---

[2] Primary field is a local operator in a conformal field theory which is annihilated by the part of the conformal algebra consisting of the lowering generators. From the representation theory point of view, a primary is the lowest dimension operator in a given representation of the conformal algebra.

where the fusion coefficients $N_{j_1,j_2,J} = 1$, $if\ J = \{|j_1 - j_2|, \ldots, min(j_1 + j_2, k - j_1 - j_2)\}$. Here $N_{j_1,j_2,J}$ are integers and known as the fusion coefficients.

7.2.5 Hilbert Space for Flat Connections

Witten established the relation that the Hilbert space for quantizing the Chern-Simons theory on M is provided by the conformal blocks of the two dimensional boundary ∂M theory. When Wilson line ends on the boundary one obtains the corresponding primary field Φ_j as a puncture. This gives the state or a conformal block. Hence the CS functional integral with the boundary ∂M is a state vector in the Hilbert space $\mathcal{H}_\Sigma$.

Hilbert Space with $\partial M = S^2$

It is easy to understand the Hilbert space for a small number of punctures. This is determined by counting the number of one-dimensional (scalar) representations that will be obtained by the tensor product of primary fields associated with the punctures.

1. If there are no punctures, then the Hilbert space is one dimensional.
2. There cannot be any non-trivial Hilbert space if there is only one puncture.
3. For two punctures with representations R_1, R_2 again the space is one dimensional if $R_1 = R_2$ for group $SU(2)$. We should have for $SU(N)$, $N \geq 3$, $R_2 = \bar{R}_1$.
4. If there are three punctures given by representations $R_i, i = 1, 2, 3$ then the dimension is given by N_{ijk}, the fusion coefficients are given by Verlinde formula [13]
5. If there are four punctures with same representations R satisfying $R \otimes R = \oplus_1^s R_i$ then it is s-dimensional for $SU(2)$.

Partition Function $Z(S^3)$

We now explain how the previous sections can be used to obtain partition function on 3-manifolds which are themselves topological invariants.

Consider a three-manifold M without any Wilson lines joined by two oppositely oriented two spheres S^2 as shown. Since the Hilbert space is one dimensional without any puncture we get the two states $|\psi\rangle$ and $|\chi\rangle$. Since the two boundaries are oppositely oriented, the partition function on M is given by the inner product:

$$Z(M) = \langle \chi | \psi \rangle. \tag{7.32}$$

We have to normalise the states with some known manifold e.g. S^3. We consider the same procedure with S^3 as the union of two balls B^3 with $\partial B^3 = S^2$ (with opposite

7.2 Chern Simons Theory and Knot/Link Invariants

orientation). Since there are no punctures again on $\mathbf{S}^2$ the states now $|v\rangle, |v'\rangle$ must be proportional to $|\psi\rangle, |\chi\rangle$. Hence we have

$$\langle \chi|\psi\rangle \langle v'|v\rangle = \langle \chi|v'\rangle \langle v|\psi\rangle$$

$$Z(M) Z(\mathbf{S}^3) = Z(M_1) Z(M_2)$$

$$\frac{Z(M)}{Z(\mathbf{S}^3)} = \frac{Z(M_1)}{Z(\mathbf{S}^3)} \frac{Z(M_2)}{Z(\mathbf{S}^3)}.$$

We can extend the above computation to the expectation value of Wilson loops. Suppose we have link L made up of n unlinked curves C_i, then we can use the above result by cutting $\mathbf{S}^3$ into balls B^3 the expectation value of the link $< V_R(L) >$ is:

$$V_R[L] = \frac{\mathcal{Z}(S^3, C_1, C_2, ..C_n)}{\mathcal{Z}(S^3)} = \prod_1^n \frac{\mathcal{Z}(S^3, C_i)}{\mathcal{Z}(S^3)}. \tag{7.33}$$

This is nothing but the result we get for polynomial invariants like Jones or HOMFLY-PT with the fundamental representation of $SU(2)$ or $SU(N)$ on the knots. The invariant for the n-component link, with the component curves C_i's carrying same representation R, is given by

$$\langle W_R(C_1, C_2, ..C_n)\rangle = \prod_1^n \langle W_R(C_i)\rangle. \tag{7.34}$$

Theorem 7.1 *For the union of disjoint oriented links L_1, L_2 with different representations R_1, R_2, the invariant is*

$$V_{R_1 R_2}[L] = V_{R_1}[L_1] V_{R_2}[L_2], \quad L = L_1 \cup L_2. \tag{7.35}$$

Consider the connected sum of two oriented links The strands that are joined have the same representation with orientations matching. Then:

Theorem 7.2

$$V_R[L_1 \# L_2] = \frac{V_R[L_1] V_R[L_2]}{V_R(\bigcirc)}, \quad \bigcirc = unknot. \tag{7.36}$$

The most important result due to Witten using this connection of Chern-Simons theory based on gauge group $SU(2)$ on $\mathbf{S}^3$ is

1. when the fundamental representation lives on all the component knots of any link, the invariant matches the Jones polynomial where $q = \exp(2\pi i/(k+2))$.
2. when the group is substituted by $SU(N)$ with fundamental representation living on all the components, we get the HOMFLY-PT two-variable polynomial. We will now derive the skein relation (7.3) within CS theory to give the three-dimensional definition of the Jones polynomial.

7.3 Jones Polynomial from CS Theory

Consider a three-manifold $\mathbf{S}^3$ with the Wilson loops carrying the fundamental representation- namely., spin-1/2 of $SU(2)$ ($R = \square$ in the Young diagram presentation) associated with any knot/link. Cut the $\mathbf{S}^3$ manifold in two parts M_L, M_R such that when we glue them back we get the manifold $\mathbf{S}^3$. Let the boundary of M_L, M_R be oppositely oriented $\mathbf{S}^2$ with four marked points (punctures) as shown in Fig. 7.5. Since $\square \times \square$ contains irreducible representations triplet $\square\!\square$ and a singlet, the Hilbert space of flat connections of the boundaries are two dimensional.

As remarked earlier the functional integral on M_L, M_R gives a state $|\chi\rangle$ and a dual vector $|\psi\rangle$. The expectation value is given by the inner product of these vectors:

$$V[L] = \langle \chi | \psi \rangle . \quad (7.37)$$

To compute the explicit polynomial form, we need to express our state $\psi\rangle$ as linear combination of the conformal block basis states, which in this case is two dimensional. Hence we have a relation:

$$|\psi\rangle + \alpha|\psi_1\rangle + \beta|\psi_2\rangle = 0 . \quad (7.38)$$

Here α, β are two complex numbers and $|\psi_1\rangle, \psi_2\rangle$ represent two different three-balls B^3 with $\mathbf{S}^2$ boundaries and four punctures each. It is seen explicitly that the three B^3 manifolds differ by the structure of the strands inside. That is., overcrossing, undercrossing or no crossing. Hence we will get the relation between

Fig. 7.5 Manifold M_L and M_R with four-punctured $\mathbf{S}^2$ boundary

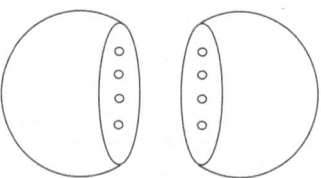

7.3 Jones Polynomial from CS Theory

three different links by taking the inner product with $|\chi\rangle$. This is equivalent to gluing M_L and M_R. Hence we have:

$$\langle\chi|\psi\rangle + \alpha\langle\chi|\psi_1\rangle + \beta\langle\chi|\psi_2\rangle = 0. \tag{7.39}$$

Thus, we obtain by gluing M_L with three different B^3 each with four punctures three different links and their polynomials. We label them as L_+, L_-, L_0. Hence we expect a recursion (skein) relation between three links obtained by shifting overcrossing to undercrossing and no crossing. Next, we have to fix the coefficients α, β from the CS theory or WZW model on the S^2 boundary. For this purpose, we understand that $|\psi_1\rangle$ is obtained from $|\psi\rangle$ by twisting the strands. Under this operation, through operator B two punctures are interchanged. Repeating once again we get the state $|\psi_2\rangle$. Hence we can write

$$|\psi_1\rangle = B|\psi\rangle, \quad |\psi_2\rangle = B^2|\psi\rangle. \tag{7.40}$$

Since B is a 2×2 matrix we can write the characteristic equation: as

$$B^2 - Tr(B) B + (Det B) = 0. \tag{7.41}$$

This immediately gives the equation we are looking for.,

$$(Det B)|\psi\rangle - Tr(B)|\psi_1\rangle + |\psi_2\rangle = 0. \tag{7.42}$$

If we know the eigenvalues of the matrix B we can get the determinant and the trace. The CFT gives the eigenvalues $\lambda_E(R, R)$ of the matrices B which is known as the monodromy matrix. Here the tensor product $\square \times \square$ gives triplet under symmetric exchange ($\square\square$) and singlet under antisymmetric exchange. The eigenvalues λ_i are:

$$\lambda_i = \pm \exp\{i\pi(2h_R - h_{R_i})\}. \tag{7.43}$$

Here $\pm$ is used for symmetric and antisymmetric products. Also, R_i are the irreducible representations of the product. The conformal weight h_R [5] is given by:

$$h_R = \frac{C_R}{k + C_V},$$

$$h_\square = \frac{3}{4(k+2)}, \quad h_{\square\square} = \frac{2}{(k+2)}, \quad h_{singlet} = 0,$$

where C_R is the Casimir invariant of the representation. Hence the eigenvalues λ_i are:

$$\lambda_{\square\square} \equiv \lambda_1 = \exp\left(\frac{-i\pi}{2(k+2)}\right), \quad \lambda_{singlet} \equiv \lambda_2 = -\exp\left(\frac{3i\pi}{2(k+2)}\right). \tag{7.44}$$

While the computations so far is correct, we need to connect with the knot theory. That requires a correction to make sure the self linking number for any knot is zero. This means that the framing should be corrected when we do the twist. This is done by multiplying the previous eigenvalues by the corresponding factor $e^{2\pi i h_\square}$. Hence the corrected eigenvalues are:

$$\bar{\lambda}_i = \exp(2i\pi h_\square)\lambda_i = \pm \exp\{i\pi(4h_\square - h_{R_i})\}. \tag{7.45}$$

This leads to $Tr(\bar{B}) = \bar{\lambda}_1 + \bar{\lambda}_2$ and $Det \bar{B} = \bar{\lambda}_1 \bar{\lambda}_2$. Now this gives us the skein relation:

$$q^{-1}V_\square[L_-] - qV_\square[L_+] = \left(q^{-\frac{1}{2}} - q^{\frac{1}{2}}\right)V_\square[L_0] \tag{7.46}$$

where $q = \exp\left(\frac{2i\pi}{k+2}\right)$. The above relation is exactly the skein relation for Jones polynomial.

There are several interesting points to be made:

1. While Chern-Simon's theory on S^3 with fundamental representation living on the links reproduces all of Jones invariants for the knots, the content of CS theory is much more.
2. The higher spin representations can live on the components of the links. These provide new polynomial invariants called colored Jones polynomials in the literature. Some of these colored Jones polynomials can distinguish inequivalent knots where Jones polynomial fails.
3. If the compact semi-simple Lie group $SU(N)$ is used along with the fundamental representation living on the components of a link, then we get the skein relation for the two-variable HOMFLY-PT Polynomial.
4. Again the higher representations lead to new invariants. But, the method of obtaining skein relation is not useful in computing these invariants. For these we refer to the papers in the literature [14].
5. The computations are conveniently made using the quantum deformations of the group $SU(2)$ known as $SU(2)_q$.
6. We also would like to point out one crucial difference. In the Jones polynomial the variable 'q' is taken to be real. But in the CS theory, the variable 'q' is a root of unity.
7. The partition function of the CS theory reflects the topological properties of the three manifolds. Different three manifolds will be characterised by these partition functions. There exists a method to compute the partition function invariant using an operation known as surgery. We will describe this in Chap. 8.

7.3.1 Wess-Zumino Witten Models, Quantum Groups, Vertex Models

Now we will describe other methods of obtaining knot/link invariants inspired by statistical mechanics. Wess-Zumino Witten (WZW) models are closely related to quantum groups. These are algebraic structures where Lie algebra is deformed in such a way the representation theory can be constructed and the new relations appear in statistical mechanics systems which are integrable. We will consider $SU(2)_q$ as an example for simplicity.

The Lie algebra of $SU(2)$ is provided by three generators J_+, J_-, J_3. They obey the commutation relations

$$[J_3, J_\pm] = \pm J_\pm,$$
$$[J_+, J_-] = J_3.$$

The q-deformed algebra is given by:

$$[J_\pm, J_3] = \pm J_\pm, \tag{7.47}$$
$$[J_+, J_-] = \frac{q^{J_3/2} - q^{-J_3/2}}{q^{1/2} - q^{-1/2}}. \tag{7.48}$$

This new algebra is no longer a Lie algebra (as the term in the RHS) is not linear). We get the algebra of $SU(2)$ in the limit $q \to 1$. The algebra Eq. (7.48) is known as $SU(2)_q$ and is a deformation of the enveloping algebra of $SU(2)$.

As the structure of the RHS will emerge everywhere it is convenient to define q-number corresponding to an integer N as:

$$[N] = \frac{q^{N/2} - q^{-N/2}}{q^{1/2} - q^{-1/2}}$$
$$= q^{(N-1)/2} + q^{(N-3)/2} + \cdots + q^{-(N-1)/2}.$$

We also define q-factorial

$$[N]! = [N][N-1][N-2]\cdots[1]. \tag{7.49}$$

These q-numbers have interesting properties:

1. $[M] \oplus [N] \equiv q^{-N/2}[M] + q^{M/2}[N] = [M+N]$,
2. $[-N] = -[N]$, $[0] = 0$, $[1] = 1$,
3. $[M][N] = [M-N] \oplus [M-N+1] \oplus \cdots [M+N]$.

In classical and quantum physics, we add angular momenta to get the total angular momentum. i.e., $\vec{J}_{total} = \vec{J}_1 + \vec{J}_2$. The basis states are $|\psi\rangle_1 \otimes \psi\rangle_2$ and the total angular momentum acts on this tensor product space. We write this explicitly as:

$$\vec{J}_{total} = \vec{J}^1 \otimes I + I \otimes \vec{J}^2. \tag{7.50}$$

This action is known as comultiplication defined as:

$$\Delta(\vec{J}) = \vec{J} \otimes I + I \otimes \vec{J}. \tag{7.51}$$

For $SU(2)_q$, the comultiplication gets modified to

$$\Delta(J_\pm) = q^{-J_3/2} \otimes J_\pm + J_\pm \otimes q^{J_3/2},$$
$$\Delta(J_3) = J_3 \otimes I + I \otimes J_3.$$

One can see the $SU(2)_q$ algebra has a $q \to q^{-1}$ symmetry, whereas the comultiplication rule breaks that. Because of this, comultiplication is noncommutative. The addition of 'q-angular momenta' depends on the order. This introduces braiding!. The representations of $SU(2)_q$ are provided by an integer or half-integer j and have dimension $2j + 1$. The representation is:

$$J_3|j, m\rangle_q = m |j, m\rangle_q,$$
$$J_\pm|j, m\rangle_q = \sqrt{[j \mp m][j \pm m + 1]}|j, m \pm 1\rangle_q.$$

There is one-to-one correspondence between representations of $SU(2)_q$ and $SU(2)$ when q is real. But when it is given as the root of unity there are several differences. For example, $q^{k+2} = 1$ as in the case of CS theory leads to the number of primary fields Φ_j being finite with $j \leq \frac{k}{2}$.

More than the dimension of the vector space, it is useful to define a new q-dimension, $\text{Dim}_q(j)$ of the representation $|j, m\rangle_q$ as:

$$\text{Dim}_q(j) = {}_q\langle j, m|Tr(q^{J_3})|j, m\rangle_q = [2j + 1]. \tag{7.52}$$

The algebraic structure of the primary fields of $SU(2)_k$ level k WZW models are given by representations of the $SU(2)_q$ where $q = \exp\left(\frac{2i\pi}{k+2}\right)$. The coproduct rules of $SU(2)_q$ apply when we take the product of representations. The states of the product representations are also related to the tensor product states by generalization of Clebsch Gordon coefficients known as q-CG coefficients. The Racah coefficients do also generalize to q-Racah coefficients

7.3 Jones Polynomial from CS Theory

For example, product of two fundamental representations $\square$ give singlet and $\square\square$. The CG coefficients relating spin-1 $|1, m\rangle$ and the two spin-1/2 states $|1/2, m_1\rangle$, $|1/2, m_2\rangle$ are:

$$|1, 1\rangle = |1/2, 1/2\rangle \, |1/2, 1/2\rangle ,$$

$$|1, 0\rangle = \frac{1}{\sqrt{2}} \left(|1/2, 1/2\rangle \, |1/2 - 1/2\rangle + |1/2, -1/2\rangle \, |1/2, 1/2\rangle \right) ,$$

$$|1, -1\rangle = |1/2, -1/2|\rangle \, |1/2, -1/2\rangle .$$

This gets modified to the form:

$$|1, 1\rangle_q = |1/2, 1/2\rangle_q \, |1/2, 1/2\rangle_q ,$$

$$|1, 0\rangle_q = \frac{q^{1/2}}{\sqrt{2}} \left(|1/2, 1/2\rangle_q \, |1/2 - 1/2\rangle_q \right)$$

$$+ \frac{q^{-1/2}}{\sqrt{2}} \left(|1/2, -1/2\rangle_q \, |1/2, 1/2\rangle_q \right) ,$$

$$|1, -1\rangle_q = |1/2, -1/2|\rangle_q \, |1/2, -1/2\rangle_q .$$

The quantum Racah coefficients for $SU(2)_q$ are given by

$$\begin{pmatrix} j_1 & j_2 & j_{12} \\ j_3 & j_4 & j_{23} \end{pmatrix} = \Delta(j_1, j_2, j_{12}) \Delta(j_3, j_4, j_{12}) \Delta(j_1, j_4, j_{23}) \Delta(j_3, j_2, j_{23})$$

$$\sum_{m \geq 0} (-)^m [m+1]! \Big\{ [m - j_1 - j_2 - j_{12}]!$$

$$[m - j_3 - j_4 - j_{12}]! [m - j_1 - j_4 - j_{23}]!$$

$$[m - j_3 - j_2 - j_{23}]! [j_1 + j_2 + j_3 + j_4 - m]!$$

$$[j_1 + j_3 + j_{12} + j_{23} - m]! [j_2 + j_4 + j_{12} + j_{23} - m]! \Big\}^{-1}$$

The $\Delta(a, b, c)$ is defined as:

$$\Delta(a, b, c) = \sqrt{\frac{[-a+b+c]![a-b+c]![a+b-c]!}{[a+b+c+1]!}} \qquad (7.53)$$

Here $[a]! = [a][a-1][a-2]\ldots[2][1]$. The $SU(2)$ spins are related as

$$\vec{j}_1 + \vec{j}_2 + \vec{j}_3 = \vec{j}_4, \quad \vec{j}_1 + \vec{j}_2 = \vec{j}_{12}, \quad \vec{j}_2 + \vec{j}_3 = \vec{j}_{23}$$

7.3.2 Vertex Models and Knot Invariants

We mentioned how the braids are related to knots. Hence this relation will also provide knot invariants from representations of braid group. If statistical mechanics models contain an R matrix which obeys Yang Baxter equation, then they are integrable. R matrix provides us with a representation of braid. Hence we can construct knot invariants through this procedure too. Jones obtained his polynomial using the integrable vertex model. We will explain the connection between the R matrix and the braid group now.

The two ends of the braids are tied together we get a knot or link. This process of tying the braid is known as closure. Alexander established a theorem that any link can be represented as the closure of a braid. However, this connection is not unique. There are two moves of braids whose closure gives the same link. These are known as Markov's moves. These can be represented through the diagram (Figs. 7.6 and 7.7).

1. Markov move 1. Consider two braids A and B. As braids AB is different from BA. However, the closure of AB and BA gives the same knot. If we denote closure of a braid A by $\widehat{A}$ then

$$\widehat{AB} = \widehat{BA}.$$

2. Markov move 2. Adding an extra braid at the end and closure gives the same knot. This is:

$$\widehat{A} = \widehat{Ab_n^{\pm}}.$$

Fig. 7.6 Markov move 1

Fig. 7.7 Markov move 2

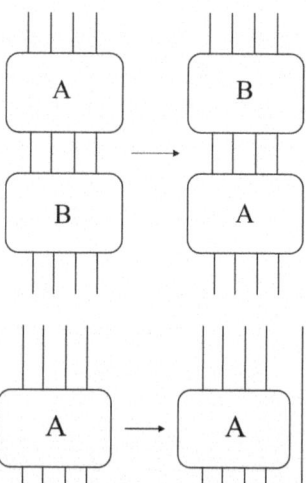

7.3 Jones Polynomial from CS Theory

7.3.3 Knot Invariant from Vertex Model

Hence to construct invariant we need to obtain a representation of the braid group. Then closure of the braid can be obtained by defining a trace operation on the braids. But, the constructed formula involving trace must remain the same under Markov moves.

We consider a vertex model as two dimensional lattice with spins on the vertices. There are four edges with each vertex. Associated with each vertex is a Boltzmann weight $S_{kl}^{ij}(u)$. Here u is known as a spectral parameter. If spin-1/2 is attached to the vertex and spin conservation is imposed at each vertex then we get six parameters S_{kl}^{ij}. This is known as the six vertex model. If spin-1 is at each vertex, then we will get 19 non-zero Boltzmann weights defining 19-vertex model. Along a row of vertices, we have the Boltzmann weight represented by

$$X_i(u) = \sum S_{lp}^{km} I^{(1)} \otimes I^{(2)} \otimes \cdots E_{pk}^{(i)} \otimes E_{ml}^{(i+1)} \otimes I^{(i+2)} \cdots I^{(n)}, \qquad (7.54)$$

where I represents identity operation and $(E_{kl})_{mp} = \delta_{km}\delta_{lp}$. The integrability of this model is shown to satisfy the Yang-Baxter equation:

$$X_i(u) X_{i+1}(u+v) X_i(v) = X_{i+1}(v) X_i(u+v) X_{i+1}(u) \qquad (7.55)$$

Interestingly the limit of spectral parameter $u \to \infty$ we get X_i satisfying braid group relation. Using this remarkable connection to the vertex model, we get Jones Polynomial from the 6 vertex model (using spin-1/2) and a new polynomial for the 19-vertex model (using spin-1). The second polynomial invariant is also obtained from Chern-Simons theory for gauge group $SU(2)$ using spin-1 living on the links of the Wilson loops.

7.3.4 Composite Braids and Murakami Invariants

There is also another way, known as composite braids which can be used to construct new knot invariants.

In Murakami's construction [15] of composite braids using r-parallel version, we replace every strand by a composite of r strands. The generators b_i, $i = 1, 2, .., n-1$ of the braid group B_n get replaced by $(n-1)$ composite braid generators $\phi^r(b_i) \in B_{rn}$ which is a map from $B_n \to B_{rn}$. In terms of the elementary braid generators b_i the composite braid generators are given by:

$$\phi^r(b_i) = (b(ri - r + 1, ri - 1))^{-r} b(ri, ri + r - 1)$$
$$b(ri - 1, ri + r - 2) \cdots b(ri - r + 1, ri).$$

where $b(i, j) = b_i b_{i+1} \cdots b_j$. We shall use a modified version of composite braids [16] where it is symmetric in the two composite legs, and each leg is twisted around

Fig. 7.8 Composite Braid
$B^{(2)}(b_1)$

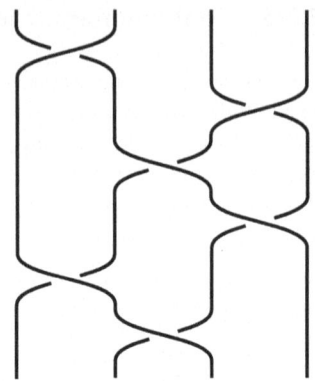

itself by π. These generators for parallel composite braids can be represented in terms of elementary braid generators as:

$$B^{(r)}(b_i) = b(ri-r+1, ri-1)^{-1} b(ri-r+2, ri-1)^{-1} \cdots b(ri-1, ri-1)^{-1}$$
$$b(ri+1, ri+r-1)^{-1} b(ri+2, ri+r-1)^{-1} .. b(ri+r-1, ri-1)^{-1}$$
$$b(ri, ri+r-1) b(ri-1, ri+r-2) \cdots b(ri-r+1, ri).$$

This we explain for $r = 2$.

$$B^{(2)}(b_i) = b_{2i-1}^{-1} b_{2i+1}^{-1} b_{2i} b_{2i+1} , b_{2i-1} b_{2i} \qquad (7.56)$$

mapping between b_i and $B^2(b_i)$ can be pictorially given as Fig. 7.8.

Similarly, the generators for the antiparallel composite braids are (figure below):

$$\tilde{B}^{(2)}(b_i) = b_{2i-1} b_{2i+1} b_{2i} b_{2i+1} b_{2i-1} b_{2i} : \qquad (7.57)$$

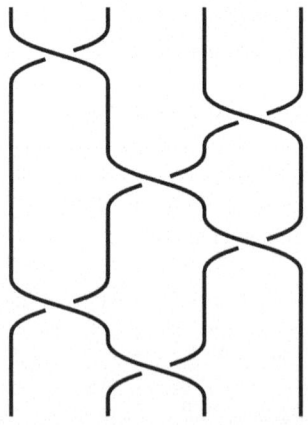

The new generators $B^{(r)}(b_i)$ satisfy the same braid group algebra, which is:

$$B^{(r)}(b_i)B^{(r)}(b_j) = B^{(r)}(b_j)B^{(r)}(b_i) \quad |i-j| \geq 2,$$
$$B^{(r)}(b_i)B^{(r)}(b_{i+1})B^{(r)}(b_i) = B^{(r)}(b_{i+1})B^{(r)}(b_i)B^{(r)}(b_{i+1}).$$

The closure of composite braids gives knots and links. Also, they respect Markov moves.

The representations of braid generators, for example, $B^{(2)}(b_i)$ can be obtained in terms of the representations of the elementary braiding generators b_i and the duality matrix relating the eigen bases of b_i and b_{i+1} in the $SU(2)$ Chern-Simons framework. We can determine the theory of composite braids for the $r = 2$ case. The generalisation to r-composite braids with an arbitrary number 'r' of strands can be easily done.

This can be extended to any compact semisimple algebra too. We will not elaborate beyond this.

Exercises

7.1. Work out the Alexander-Conway polynomial for the figure eight (knot 4_1).

7.2. Verify the Jones polynomial for trefoil T.

7.3. Obtain HOMFLY-PT polynomial for trefoil T and figure 8 (knot 4_1).

7.4. Work out the Kauffman polynomial for trefoil T.

7.5. Verify that the solution for $F = dA + A \wedge A = 0$ is $A = g^{-1}dg$.

7.6. If spin-1 states are considered in a vertex model, list the number of number of non-zero Boltzmann weights.

References

1. J. Baez, *Knots and Quantum Gravity* (Oxford University Press, 1994)
2. E. Artin, Ann. Math. **48**, 101 (1947)
3. F. Wilzcek, Phys. Rev. Lett. **49**, 957 (1982)
4. A.B. Zamolodchikov, Conformal symmetry in two-dimensional space: Recursion representation of conformal block. Theor. Math. Phys. **73**, 4 (1988)
5. E. Witten, Quantum field theory and Jones polynomial. Commun. Math. Phys. **121**, 351 (1989)
6. W. Thomson, On vortex atoms. Proc. R. Soc. Edinb. **6**, 94 (1867)
7. L. Kauffman, An invariant of regular isotopy. Trans. Am. Math. Soc. **318**, 417 (1990)
8. P. Freyd, D. Yetter, J. Hoste, W.B.R. Lickorish, K. Millett, A. Ocneanu, A new polynomial invariant of knots and links. Bull. Am. Math. Soc. **12**, 239 (1985)

9. J.H. Przytycki, P. Traczyk, Invariants of links of Conway type. Kobe. J. Math. **4**, 115 (1987); arXiv:1610.06679
10. R.K. Kaul, T.R. Govindarajan, P. Ramadevi, Schwarz type topological field theories, in *Encyclopedia of Mathematical Physics*, ed. by J.P. Francoise, G.L. Naber, T.S. Tsun
11. V.G. Kac, Simple irreducible graded Lie algebras of finite growth. Math. USSR Izv. **2**, 1271 (1968); R.V. Moody, A new class of Lie algebras, J. Algebra, 10, 211 (1968)
12. H. Sugawara, A field theory of currents. Phys. Rev. **170**, 1659 (1968)
13. E. Verlinde, Fusion rules and modular transformations in 2D conformal field theory. Nuclear Phys. B **300**, 360 (1988)
14. R.K. Kaul, T.R. Govindarajan, Three dimensional Chern-Simons theory as theory knots and links. Nuc. Phys. **B380**, 293 (1992); P. Rama Devi, T.R. Govindarajan, R.K. Kaul, Three dimensional Chern-Simons theory as theory knots and links III: Compact semi-simple group. Nuc. Phys. B402, 548 (1993)
15. J. Murakami, Osaka J. Math. **26**, 1 (1989)
16. P. Ramadevi, T.R. Govindarajan, R.K. Kaul, Mod. Phys. Lett. A10, 1635 (1995)

Three-Manifold Invariants

We know that the Euler characteristic χ completely classifies topological structure of the two-dimensional surfaces: two-sphere $\mathbf{S}^2$, torus $\mathbf{T}^2$ and higher genus Riemann surfaces Σ_g as pointed out in the earlier chapters. But an analogous mathematical quantity is not available to distinguish different three-dimensional manifolds. Some of the well-known three-manifolds are three-sphere $\mathbf{S}^3$, $\mathbf{S}^2 \times \mathbf{S}^1$, $\mathbf{T}^3$. Besides these, there are quotienting of the three-sphere

(i) lens spaces $L(p,1) \equiv \mathbf{S}^3/\mathbb{Z}_p$ (and general $L(p,q)$ with p,q coprime) where the complex coordinates, obeying $|z_1|^2 + |z_2^2| = $ const satisfy the following equivalence relation:

$$(z_1, z_2) \sim (e^{2\pi i/p} z_1, e^{-2\pi i q/p} z_2),$$

(ii) $\mathbf{S}^3/G$ (G any discrete subgroup of $SU(2)$) Poincare manifold $\mathbf{P}^3$ and also hyperbolic manifolds H^3 defined in Chap. 6.

Every three-manifold can be constructed through a procedure called *surgery* of knots and links in $\mathbf{S}^3$ (Lickorish-Wallace theorem [1]). Conversely, the surgery of knots and links in any three-manifold can lead to the three-sphere $\mathbf{S}^3$.

We will briefly review the surgery procedure in the following section. In the subsequent sections, we construct the three-manifold invariants [2] using the $SU(2)$ Chern-Simons theory invariants for the knots and links [3–5] discussed in the previous chapter.

8.1 Surgery Procedure

We will now illustrate the steps constituting the surgery on a mathematical circle C inside a three-manifold $M = \mathbf{S}^2 \times \mathbf{S}^1$ (see Fig. 8.1):
(For clarity, let us take C to be a simple unknot.)

(i) We thicken C by taking the tubular neighbourhood $\mathbf{N}$ as indicated in Fig. 8.2.
(ii) Remove $\mathbf{N}$ which is a solid torus whose boundary $\partial \mathbf{N} = \mathbf{T}^2$ (two torus with a and b cycles referred to as a meridian and a longitude, remember $\pi_1(\mathbf{T}^2 = \mathcal{Z} \oplus \mathcal{Z})$ as shown in Fig. 8.3.
(iii) The remainder $\mathbf{S}^2 \times \mathbf{S}^1 - \bar{\mathbf{N}}$, where $\bar{\mathbf{N}}$ is the interior of $\mathbf{N}$, is also a solid torus with oppositely oriented $\mathbf{T}^2$ boundary. The remainder $\mathbf{S}^2 \times \mathbf{S}^1 - \bar{\mathbf{N}}$ is also referred to as the complement $(\mathbf{S}^2 \times \mathbf{S}^1)/C$ in the literature and is depicted in Fig. 8.4.
(iv) Gluing $\mathbf{N}$ with $(\mathbf{S}^2 \times \mathbf{S}^1)/C$ by identifying the points on the boundary through the homeomorphism:

$$h : \partial \mathbf{N} \to \partial \left[(\mathbf{S}^2 \times \mathbf{S}^1)/C \right] \tag{8.1}$$

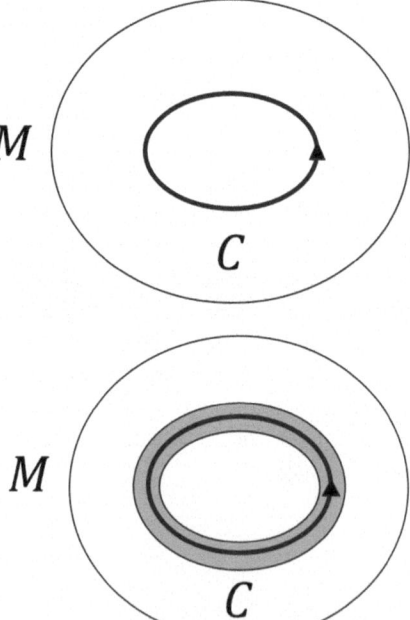

Fig. 8.1 Mathematical curve C in $M = \mathbf{S}^2 \times \mathbf{S}^1$

Fig. 8.2 Tubular neighbourhood of C in $M = \mathbf{S}^2 \times \mathbf{S}^1$.

8.1 Surgery Procedure

Fig. 8.3 Removed tubular Neighbourhood **N** with T^2 boundary

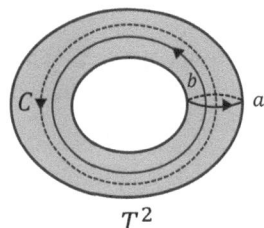

Fig. 8.4 Knot complement $M/C = (S^2 \times S^1)/C$ with T^2 boundary

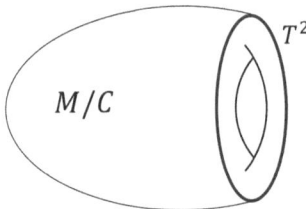

leads to a new three-manifold M which depends on C as well as the homeomorphism:

$$\hat{M} \equiv (S^2 \times S^1)/C \bigcup_h N .$$

(i) When $h = \mathbb{I}$ (identity operation), it is straightforward to visualise that gluing the two solid tori **N** with $(S^2 \times S^1)/C$ gives back the three-manifold $\hat{M} \equiv (S^2 \times S^1)$.

(ii) Suppose the homeomorphism maps meridian a in Fig. 8.3 to longitude b in Fig. 8.4 and vice-versa:

$$h : a \to b, \ b \to -a ,$$

then gluing will result in a topologically different three-manifold $\hat{M}$.

For C taken as unknot, this simple map can be viewed as an inversion on a solid torus resulting in $(\mathbb{R}^3 \bigcup \infty)/C$. Hence gluing $(\mathbb{R}^3 \bigcup \infty)/C \bigcup N$ leads to $(\mathbb{R}^3 \bigcup \infty)$ which is topologically equivalent to the three-sphere S^3. Such a visualisation of $\hat{M}$ will not be easy for a general curve C and homeomorphism h.

Instead of starting from knots or links in $S^2 \times S^1$, the three-manifolds can also be constructed from surgery of knots or links inside S^3 (Lickorish Wallace theorem) [1].

(continued)

Theorem 8.1 (Lickorish-Wallace Theorem) *Any connected closed orientable three manifold can be obtained by surgery on a framed link in S^3.*

Even though we have elaborated surgery on a knot, the procedure can be performed on any r-component link as well. Taking a tubular neighbourhood of such links will lead to a three-manifold with r disjoint $\mathbf{T}^2$ boundaries. Then, the homeomorphic map can be done on each of the torus boundaries and then glued with the link complement to obtain a new three-manifold. This is referred to in the literature as *Dehn filling* for surgery of knots/links in S^3.

There is another subtle point about knots/links which we would like to emphasise in the context of surgery. We saw in the earlier chapter that any two knots/links are equivalent if we can go from one to another under Reidmeister moves I, II and III. Such knots/links are said to be *ambient isotopic* to each other. However, surgery of any of the two knots/links equivalent under Reidmeister moves II and III will give topologically the same three-manifold. Such knots/links are called *framed knots/links*. Two framed knots related by Reidmeister moves II and III are said to be *regular isotopic* to each other.

The crossing numbers f_i's of the component knots $\mathcal{K}_i$'s constituting a r-component framed link $[\mathcal{L}, \mathbf{f}]$ are referred to as framing numbers. Here the boldfaced $\mathbf{f} = (f_1, f_2 \ldots f_r)$. For example, unknot U with framing number $f = 0$ is $[U, 0] = \bigcirc$. Adding f number of righthanded half-twists will give unknot with framing $+f$.

We tabulate a list of knots and/or links in Fig. 8.5 along with the corresponding three-manifold [1] under homeomorphic mapping $h : a \to b$. From the tabulation, we see that the surgery of one or more framed knots/links gives the same three-manifold. That is, the map from framed links to three-manifolds is not one-to-one and onto.

Framed Link Diagram	3-manifold	Framed Link Diagram	3-manifold
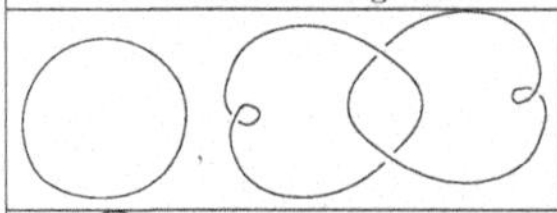	$S^2 \times S^1$	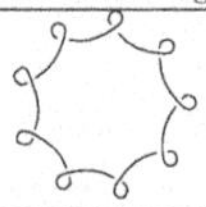	$L(9,1)$
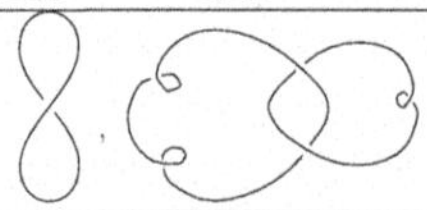	S^3	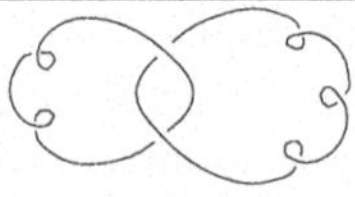	$L(5,1)$
	$\mathbb{RP}^3$		$\mathbb{P}^3$

Fig. 8.5 Framed link(s) and the corresponding three-manifold

8.2 Kirby Moves

Analogous to the Markov moves, there are two Kirby moves [6] on any framed link which gives topologically the same three-manifold. The two Kirby moves on a framed knot $U(Y)$ are shown in Fig. 8.6. Note that the Y denotes a non-trivial region with self-crossings resulting in the framed knot $U(Y)$. Under Kirby move I, we can see that the framed knot $U(Y)$ transforms to a two-component framed link $H(Y)$. Note that there is a curve C with the framing number $+1$ going around the framed knot with additional twisting.

For any general r-component framed link $[\mathcal{L}, \mathbf{f}]$, the Kirby move I will convert it to an $r+1$-component framed link due to the curve C. The framing numbers f_j's of the component knots $\mathcal{K}_j$'s of r-component link will change to

$$f'(j) = f(j) - \bigl(Lk(\mathcal{K}_j, C)\bigr)^2, \qquad (8.2)$$

where $Lk(\mathcal{K}_j, C)(7.12)$ is the linking number of the component knot $\mathcal{K}_j$ and the closed curve C.

The Kirby move II adds a disjoint framed unknot $[U, \pm 1]$ with framing ± 1 to the framed knot $U(Y)$ or any r-component framed link $[\mathcal{L}, \mathbf{f}]$. The three-manifold obtained by surgery of $U(Y)$ will be the same as that obtained from $U(Y) \times [U, \pm 1]$. The framed unknot in the Kirby move II drawn in Fig. 8.6 is with framing -1.

Theorem 8.2 (Kirby Theorem) *Two framed links determine the same three manifold if and only if they are related by a sequence of diagram moves which are referred to as Kirby moves.*

In the following section, we would like to obtain an algebraic expression incorporating the surgery of framed links in $\mathbf{S}^3$. Such an expression will qualify as a three-manifold invariant provided it remains the same when we replace the framed link with another framed link related by Kirby moves I and II.

Fig. 8.6 Kirby moves on any framed link $U(Y)$

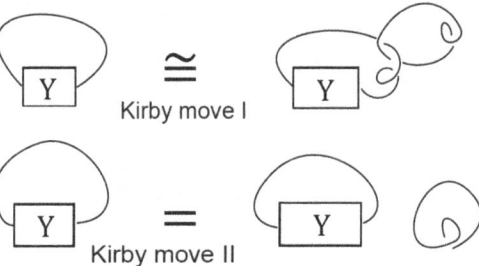

8.3 Three-Manifold Invariants

Let us focus on the construction of algebraic expression for three-manifold invariants in $SU(2)$ Chern-Simons theory with coupling constant $k = \mathbb{Z}_+$ [2].

In the previous chapter, we discussed formal integral expression of $SU(2)$ Chern-Simons partition function $Z[M]$ involving Chern-Simons action S on any three-manifold M. Using canonical quantization, the partition function for $\mathbf{S}^2 \times \mathbf{S}^1$ can be shown to be

$$Z[\mathbf{S}^2 \times \mathbf{S}^1] = \langle \chi_0 | \xi_0 \rangle = 1 . \tag{8.3}$$

The state $|\xi_0\rangle$ is same as $|\chi_0\rangle$ as the Hilbert space on $\mathbf{S}^2$ is one-dimensional. Further, inserting a Wilson line in spin representation j, the partition function in the presence of the Wilson line

$$Z[\mathbf{S}^2 \times \mathbf{S}^1; R_j] = \delta_{j,0} , \tag{8.4}$$

as the Hilbert space of the one-punctured $\mathbf{S}^2$ boundary is also one-dimensional as discussed in the previous chapter (Sect. 7.2.5). Hence, the above expression is non-zero only for trivial representation(singlet) on the Wilson line.

Using the above data and the homeomorphism map h, we can determine the partition function for the other three manifolds. As the closed curve C is only a mathematical one, it can be considered to carry a trivial representation (trivial/identity representation denoted as 0). The Chern-Simons partition function on $\hat{M}$ will be

$$Z(\hat{M}) = \sum_i h_{0i} Z(\mathbf{S}^2 \times \mathbf{S}^1; R_i) . \tag{8.5}$$

The summation is over allowed spin representations $0 \leq i \leq k/2$ of the $su(2)_k$ current algebra/Kac-Moody discussed in the previous chapter. The homeomorphism h will be a matrix in this space of representations $|R_i\rangle$'s with matrix elements $h_{ij} \equiv \langle R_i | h | R_j \rangle$.

For the three-sphere $\mathbf{S}^3$, the homeomorphism mapping the meridian to the longitude is given by the modular transformation matrix S. Hence the partition function $Z[\mathbf{S}^3]$ is given by

$$Z[\mathbf{S}^3] = \sum_j S_{0j} Z(\mathbf{S}^2 \times \mathbf{S}^1; R_j) = S_{00} , \tag{8.6}$$

where the matrix elements of the modular transformation matrix S i [3] is given by

$$S_{ij} = \sqrt{\frac{2}{k+2}} \sin \frac{\pi(2i+1)(2j+1)}{k+2} . \tag{8.7}$$

8.3 Three-Manifold Invariants

Instead of considering surgery of C in $\mathbf{S}^2 \times \mathbf{S}^1$, we could perform surgery of framed knots and links in $\mathbf{S}^3$. It is to be remembered that framed knot or link invariants will require the braiding eigenvalues without the framing corrections (Eq. (7.45)). For example invariant for unknot with framing $p : [U, p]$ will be $V_R(\bigcirc) q^{p C_R}$. So, we incorporate such a modification in the knot and link invariants discussed in the previous chapter to obtain the framed knot and framed link invariants.

Using these framed knot and link invariants, we can construct the three-manifold partition function from the surgery of these knots/links in $\mathbf{S}^3$. Algebraically, the three-manifold obtained from the surgery of a framed link $[L, \mathbf{f}]$ in $\mathbf{S}^3$ must be written in terms of linear combinations of the colored framed link invariants where the component knots carry different spins (colors) of $SU(2)$ gauge group. Such a mathematical expression will qualify as a 'three-manifold invariant' if and only if it remains unchanged for the above two Kirby moves on the framed link.

Consider a three-manifold $\hat{M}$ obtained from surgery of a r-component framed link $[L, \mathbf{f}]$ in $\mathbf{S}^3$. Then a three-manifold invariant $\hat{F}^{(SU(2))}[\hat{M}]$ for $\hat{M}$ is given as a linear combination of the colored framed link invariants $V^{SU(2)}_{R_1, R_2, \ldots R_r}[L, \mathbf{f}]$ with representations $R_1, R_2, \ldots R_r$ living on component knots is [2, 7]:

$$\hat{F}^{(SU(2))}[\hat{M}] = \alpha^{-\sigma[L,\mathbf{f}]} \sum_{R_1, R_2, \ldots R_r=0}^{k/2} \left(\prod_{i=1}^{r} \mu_{R_i} \right) V^{SU(2)}_{R_1, R_2, \ldots R_r}[L, \mathbf{f}] , \quad (8.8)$$

Here $\sigma[L, \mathbf{f}]$ is the signature of the linking matrix. Note that the range of the $SU(2)$ representations will be $R_i = [0, 1/2, \ldots k/2]$ where 0 refers to singlet and $k/2$ refers to spin $k/2$. Clearly, the final summed up quantity $\hat{F}^{(SU(2))}[\hat{M}]$ is representation independent.

Imposing the above expression to remain unchanged under Kirby moves suggests an explicit form for α and μ_{R_i}. The explicit form of framed link invariants for $U(Y)$, $H(Y)$ and $U(Y) \times U^{\pm}$ along with the knowledge of the known identities obeyed by the modular transformation matrices S, T of the $su(2)_k$ Wess-Zumino-Witten conformal field theory [8] led to the following solution for α and μ_{R_i} [7]:

$$\mu_{R_i} = S_{0i} , \quad \alpha = e^{i\pi c/4} . \quad (8.9)$$

involving the $su(2)_k$ conformal field theory central charge c and the modular transformation matrix S.

We refer interested readers to Refs. [2, 7] for proofs showing $\hat{F}^{SU(2)}[M]$ remain unchanged under both Kirby moves I and II. In Chern-Simons field theory, we expect the three-manifold invariant $\hat{F}^{SU(2)}[\hat{M}]$, which is color independent, must be proportional to the partition function $Z[\hat{M}]$ for three-manifold $\hat{M}$. The exact proportionality constant can be inferred from studying the explicit $\hat{F}^{SU(2)}[\hat{M}]$ for some three-manifolds: $\mathbf{P}^3$, $\mathbf{RP}^3$, Lens spaces $L(p, 1)$. In the mathematics literature these invariants are known as Witten-Reshetikhin-Turaev (WRT) invariants.

The algebraic expressions of this invariant calculated explicitly from the formula in Eq. (8.8) for some three manifolds are tabulated below. We have indicated the framed links in $\mathbf{S}^3$ which on surgery yields the corresponding three-manifolds $\hat{M}$. C_R is the quadratic Casimir for the representation R of the gauge group $SU(2)$.

Framed link	$\hat{M}$	$\hat{F}^{(SU(2))}[\hat{M}]$
Unknot with zero framing	$\mathbf{S}^2 \times \mathbf{S}^1$	$1/S_{00}$
Unknot with framing ± 1	$\mathbf{S}^3$	1
Unknot with framing $+2$	$\mathbf{RP}^3$	$\alpha^{-1} \sum_R \frac{S_{0R} q^{2C_R} S_{0R}}{S_{00}}$
Unknot with framing $+p$	$L[p,1]$	$\alpha^{-1} \sum_R \frac{S_{0R} q^{pC_R} S_{0R}}{S_{00}}$

$\hat{F}^{(SU(2))}[\hat{M}]$ *for some manifolds*

For the $SU(2)$ Chern-Simons theory on a three-manifold $\hat{M}$, the partition function $Z^{SU(2)}[\hat{M}]$ is also an equivalent invariant characterising the three-manifold $\hat{M}$. This has been calculated for several manifolds by different methods [9]. Comparing these expressions with the corresponding invariants $\hat{F}^{(SU(2))}[\hat{M}]$ tabulated above, it appears that the three-manifold invariant $\hat{F}^{(SU(2))}[\hat{M}]$ is related to the Chern-Simons partition function $Z^{SU(2)}[\hat{M}]$ as follows:

$$\hat{F}^{(SU(2))}[\hat{M}] = \frac{Z^{(SU(2))}[\hat{M}]}{S_{00}}. \qquad (8.10)$$

Lickorish [10] had an expression for the three-manifold invariant $F^l[M]$ as a linear combination of bracket polynomials of cables of a framed link on $\mathbf{S}^3$, which under surgery gives the three-manifold $\hat{M}$. Even though, Jones' polynomial in variable q equals bracket polynomial in variable A when $q^{1/4} = -A$, it was not clear whether $F^l[\hat{M}]$ is actually equal to $\hat{F}^{(SU(2))}[\hat{M}]$ under $q^{1/4} = -A$. Using the representation theory of composite braids [11], such a cable of any link diagram can be viewed as a tensor product of spin 1/2 representations. This enabled in proving [12]:

$$F^l[M]|_{A=-q^{1/4}} = F^{(SU(2))}[M]. \qquad (8.11)$$

Even though we focused on three-manifold invariants in $SU(2)$ Chern-Simons theory, it is a straightforward exercise to generalise to other compact semi-simple gauge group $\mathcal{G}$ [7].

8.4 Conclusion

In this chapter, we briefly reviewed the surgery prescription of obtaining other three-manifolds. Then, we have focused on the algebraic expression for the three-manifold invariants. Such an expression quantifies the Lickorish-Wallace theorem which states that three-manifolds are obtained by surgery of framed knots and links. Further, the two Kirby moves I and II on framed links do not alter the three-manifold invariants.

Using Chern-Simons field theory invariants for framed links colored by representations of the $SU(2)$ gauge group, we have discussed the construction of three-manifold invariants (8.8). Particularly, the invariant remains unaltered under the Kirby moves on framed links. We discussed $SU(2)$ gauge group for clarity in this chapter and presented three-manifold invariants for some three-manifolds.

There are mutant pairs of framed knots and links, which share the same $SU(2)$ Chern-Simons field theory invariant [13]. Hence, $SU(2)$ three-manifold invariants for mutant pairs are the same. Note that the mutant pairs of framed links are not related by the Kirby moves. With one counterexample, we can assert that these $SU(2)$ three-manifold invariants do not completely classify inequivalent three-manifolds.

The detection of mutation operation for colored framed links is now possible if we go to higher rank gauge groups like $SU(3)$, $SU(4)$ etc. Unlike the $SU(2)$ gauge group, whose representations are only totally symmetric representations, there are mixed representations for the higher rank gauge groups. These mixed representations placed on the framed link components play a crucial role in detecting mutation [14].

Hence, the invariant (8.8) for $SU(N)$ gauge group will qualify as a promising candidate for the three-manifold invariants.

Besides its applications to mathematics, Chern-Simons theory also has interesting applications in the $2+1$ dimensional black hole physics. We will discuss the salient features of black hole physics applications in the next chapter.

Exercises

8.1 Work out the invariant for trefoil with framing number $p = +3$.

8.2 Obtain the three manifold invariant Eq. (8.10) for trefoil with framing $p = +1$ as indicated in Fig. 8.5 ($\mathbf{P}^3$, Poincare manifold).

8.3 Work out three manifold invariant for unknot $[U, 0]$ and framed Hopf link $[H, \mathbf{F} = (1, 1)]$ and verify they are same.

References

1. A.D. Wallace, Can. J. Math. **12**, 503 (1960); W.B.R. Lickorish, Ann. Math. **76**, 531 (1962)
2. R.K. Kaul, Chern-Simons theory, knot invariants, vertex models and three-manifold invariants, hep-th/9804122, in *Frontiers of Field Theory, Quantum Gravity and Strings*, vol. 227, ed. by R.K. Kaul et al. Horizons in World Physics (NOVA Science Publishers, New York, 1999)
3. E. Witten, Commun. Math. Phys. **121**, 351 (1989)
4. R.K. Kaul, Topological quantum field theories - a meeting ground for physicists and mathematicians, hep-th/9907119, in *Quantum Field Theory: A 20th Century Profile*; R.K. Kaul, Complete solution of $SU(2)$ Chern-Simons theory, hep-th/9212129; R.K. Kaul, Commun. Math. Phys. **162**, 289 (1994)
5. P. Ramadevi, T.R. Govindarajan, R.K. Kaul, Nucl. Phys. **B402**, 548 (1993)
6. R. Kirby, Invent. Math. **45**, 35 (1978); R. Fenn, C. Rourke, Topology **18**, 1 (1979)
7. R.K. Kaul, P. Ramadevi, Commun. Math. Phys. **217**, 295 (2001)
8. P. Di Francesco, P. Mathieu, D. Senechal, in *Conformal Field Theory*, ed. by J.L. Birman, J.W. Lynn, M.P. Silverman, H.E. Stanley, M. Voloshin. Graduate Texts in Contemporary Physics (Springer, Berlin, 1997)
9. L.C. Jeffrey, Commun. Math. Phys. **147**, 563 (1992)
10. W.B.R. Lickorish, Three manifolds and Temperley Lieb algebra. Math. Ann. **290**, 657–670 (1991); W.B.R. Lickorish, Calculations with the Temperley-Lieb algebra. Comment. Math. Helvetici **67**, 571–591 (1992)
11. P. Ramadevi, T.R. Govindarajan, R.K. Kaul, Mod. Phys. Lett. **A10**, 1635 (1995)
12. P. Ramadevi, S. Naik, Commun. Math. Phys. **209**, 29 (2000)
13. P. Ramadevi, T.R. Govindarajan, R.K. Kaul, Mod. Phys. Lett. **A9**, 3205 (1994)
14. S. Nawata, P. Ramadevi, V.K. Singh, Colored HOMFLY polynomials that distinguish mutant knots. J. Knot Theory Ramif. **26**(14), 175009 (2017); A. Mironov, A. Morozov, An. Morozov, P. Ramadevi, V.K. Singh, Colored HOMFLY polynomials of knots presented as double fat diagrams. J. High Energy Phys. **1507**, 109 (2015); L. Bishler, S. Dhara, T. Grigoryev, A. Mironov, A. Morozov, Difference of mutant knot invariants and their differential expansion. Zh. Eksp. Teor. Fiz. **111**, N9 (2020); L. Bishler, S. Dhara, T. Grigoryev, A. Mironov, A. Morozov, An. Morozov, P. Ramadevi, V.K. Singh, A. Sleptsov, Distinguishing mutant knots. J. Geom. Phys. **159**, 103928 (2021)

3D Gravity and BTZ Blackhole

Three-dimensional gravity is an excellent model for understanding several features of topological and quantum aspects of gravity. This is because in three-dimensional gravity we do not have propagating (dynamical) degrees of freedom. But topological aspects provide interesting features. There is one more reason to understand this model. It happens that the near horizon aspects of higher dimensional black holes contain such topological information [1].

Let us recall the Riemann tensor discussed in Chap. 3. It was pointed out that the Riemann tensor measures how the space is curved. If a vector is parallelly transported along a closed loop this tensor measures the difference between the vector before and after the transport. It was also explained that this tensor $R^\rho_{\sigma\mu\nu}$ is given in terms of Christoffel connections $\Gamma^\rho_{\nu\sigma}$ by:

$$R^\rho_{\sigma\mu\nu} = \partial_\mu \Gamma^\rho_{\nu\sigma} - \partial_\nu \Gamma^\rho_{\mu\sigma} + \Gamma^\rho_{\mu\lambda}\Gamma^\lambda_{\nu\sigma} - \Gamma^\rho_{\nu\lambda}\Gamma^\lambda_{\mu\sigma}. \tag{9.1}$$

Also, let us recollect that the Christoffel connection $\Gamma^\rho_{\nu\sigma}$ is given in terms of the metric $g_{\mu\nu}$ by:

$$\Gamma^\rho_{\mu\nu} = \frac{1}{2} g^{\rho\alpha} \left(\partial_\mu g_{\alpha\nu} + \partial_\nu g_{\mu\alpha} - \partial_\alpha g_{\mu\nu} \right). \tag{9.2}$$

The trace of the Riemann curvature tensor is the Ricci tensor $R_{\mu\nu}$ and the trace of the Ricci tensor is the scalar curvature R:

$$R_{\mu\nu} = R^\rho_{\mu\rho\nu}, \quad R = R^\mu_\mu. \tag{9.3}$$

Finally the Einstein tensor is:

$$G_{\mu\nu} = R_{\mu\nu} - \frac{1}{2} g_{\mu\nu}. \tag{9.4}$$

© The Author(s), under exclusive license to Springer Nature Switzerland AG 2024
T. R. Govindarajan, P. Ramadevi, *Geometry and Topology of Low Dimensional Systems*, Lecture Notes in Physics 1027,
https://doi.org/10.1007/978-3-031-59501-1_9

We recollect that the gravity in any dimension with cosmological constant Λ in the absence of matter is described by

$$G_{\mu\nu} + \Lambda\, g_{\mu\nu} = 0. \tag{9.5}$$

When $\Lambda = 0$ we get asymptotically Minkowski space. With $\Lambda = \pm\frac{1}{l^2}$ we get anti-de Sitter or de Sitter space.

9.1 3D Gravity

It was pointed out earlier that in three dimensions the Riemann tensor $R^{\rho}_{\lambda\mu\nu}$ has six independent components. Ricci tensor also being a symmetric matrix has six components. The Weyl tensor vanishes completely [2]. The 3D gravity has no propagating degrees of freedom. But quantum theory of gravity will depend on the topological features of the three manifold on which gravity is described. It has interesting topological features which are useful in higher dimensions. In this section we will see that the topological degrees of freedom of the 3D gravity will be described by a Chern-Simons theory with appropriate group depending on the cosmological constant [2].

From Einstein's equation (Eq. 9.5) we get

$$R = 6\Lambda, \tag{9.6}$$

which leads to $R_{\mu\nu} = 2\Lambda\, g_{\mu\nu}$. This completely fixes the Riemann tensor in terms of the metric to be

$$R_{\mu\nu\rho\sigma} = \frac{R}{6}\left(g_{\mu\rho}g_{\nu\sigma} - g_{\mu\sigma}g_{\nu\rho}\right). \tag{9.7}$$

Hence Riemann tensor is given entirely by the cosmological constant Λ and the metric. Such spaces are known as maximally symmetric ones. Maximally symmetric spaces have the same number of symmetries and Killing vectors (except for the conformal ones) as Euclidean spaces and locally equivalent everywhere.

Positive sign of the cosmological constant corresponds to spherical spaces and negative constant will be hyperboloids also known as anti-de Sitter space (AdS_3). The geometry of these spaces as given by the metric can be obtained by considering the hyperboloid embedded in four dimensions:

$$x_1^2 + x_2^2 - x_3^2 - x_4^2 = \Lambda = -\frac{1}{\ell^2}. \tag{9.8}$$

The four-dimensional space has the Pseudo-Riemannian metric:

$$ds^2 = dx_1^2 + dx_2^2 - dx_3^2 - dx_4^2. \tag{9.9}$$

9.1 3D Gravity

We can parametrize the hypersurface as

$$x_1 = \frac{1}{\ell}\sinh\alpha\cos\theta, \quad x_2 = \frac{1}{\ell}\sinh\alpha\sin\theta$$
$$x_3 = \frac{1}{\ell}\cosh\alpha\cos\phi, \quad x_4 = \frac{1}{\ell}\cosh\alpha\sin\phi.$$

The metric reduces to

$$ds^2 = \frac{1}{\ell^2}\left(\sinh^2\alpha\, d\theta^2 + d\alpha^2 - \cosh^2\alpha\, d\phi^2\right). \tag{9.10}$$

Here ϕ is timelike due to the negative sign, but it is also an angle with $\phi \equiv \phi + 2\pi$. To make it a proper time coordinate we remove this condition by defining $\phi = \frac{t}{\ell}$ and $r = \frac{\ell}{\sinh\alpha}$. Hence we get the global metric for the universal covering space of AdS_3 as:

$$ds^2 = -\left(1 + \frac{r^2}{\ell^2}\right)dt^2 + \frac{1}{\left(1 + \frac{r^2}{\ell^2}\right)}dr^2 + r^2 d\theta^2. \tag{9.11}$$

While there are no blackhole solutions in zero cosmological constant spaces, they do exist in AdS spaces (i.e with negative cosmological constant). These were shown to exist in these spaces by simple group theory arguments by Banodos, Teitelboim and Zanelli [3] (BTZ).

9.1.1 Killing Symmetries of AdS_3

As pointed out earlier AdS_3 is maximally symmetric and it is easy to see from Eq. (9.8) the symmetry group as $SO(2, 2)$. The group algebra is the same as that of $SL(2, R) \otimes SL(2, R)$. $SL(2, R)$ can also be viewed as $SU(1, 1)$. This also acts on a complex upper half plane through:

$$z \to z' = \frac{az + b}{cz + d}, \quad ac - bd = 1. \tag{9.12}$$

The symmetries are provided by the Killing vectors. There are six generators of the symmetries $SL(2, R) \otimes SL(2, R)$. They are denoted by $J_\pm$, J_0 and $K_\pm$, K_0. One can construct new geometries as coset spaces by the action of the discrete subgroups of the Killing symmetries. This will provide us with new solutions which are locally the same but globally different. It was shown by BTZ [3] that such a space to be a black hole solution in 3 dimensional asymptotically Anti-de Sitter space.

The Killing symmetries correspond to time translations and rotations. Killing vector defines one parameter subgroup of isometries of the AdS space. It is defined for $P \in AdS_3$

$$P \longrightarrow e^{t\zeta} P. \tag{9.13}$$

Since the transformation Eq. (9.13) are isometries the coset space obtained by the identifications, it inherits from AdS space the constant curvature metric. But an important restriction arises from the requirement that there should not be closed timelike curves. This requires the Killing vector ζ as spacelike i.e.

$$\zeta \cdot \zeta > 0. \tag{9.14}$$

The Killing vector can be seen to be given by

$$\zeta = \frac{r_+}{\ell} J_3 - \frac{r_-}{\ell} K_3 - J_-. \tag{9.15}$$

Writing $\zeta = \frac{1}{2}\omega_{ab}J^{ab}$ we obtain the values of the Casimir invariants as:

$$2M = -\omega_{ab}\omega^{ab} = \frac{2}{\ell^2}\left(r_+^2 + r_-^2\right), \tag{9.16}$$

$$2J = -\frac{1}{2}\ell\epsilon^{abcd}\omega_{ab}\omega_{cd} = \frac{4}{\ell}r_+ r_-. \tag{9.17}$$

Here M, J are the mass and angular momentum of the blackhole given in terms of the inner and outer horizons. It is obvious that $J \leq \ell M$. The extremal case corresponds to the geometry with only one horizon or when inner and outer horizons coincide i.e., $r_+ = r_-$. This happens when J takes its maximum value $J = \ell M$.

9.2 Chern Simons Formulation of 3D Gravity

Having given the geometric formulation 3D gravity and BTZ blackhole with cosmological constant, we now argue this geometric description can be explained in terms of a topological gauge theory which we are familiar with, namely the Chern-Simons theory.

For this purpose we start with the definition of triads e^a_μ

$$e^a_\mu e^\nu_a = \delta^\nu_\mu, \quad e^a_\mu e^\mu_b = \delta^a_b \tag{9.18}$$

and

$$e^a_\mu e^b_\nu \eta_{ab} = g_{\mu\nu}. \tag{9.19}$$

9.2 Chern Simons Formulation of 3D Gravity

The covariant derivative of the tensor field, given in local coordinates as V_b^a is given in terms of spin connection as:

$$D_\mu V_b^a = \partial_\mu V_b^a + \omega_{\mu c}^a V_b^c - \omega_{\mu b}^c V_c^a. \tag{9.20}$$

But the spin connection $\omega_{\mu c}^a$ is related to the Christoffel connection through:

$$\omega_{\mu b}^a = e_\alpha^a e_b^\beta \Gamma_{\mu\beta}^\alpha - e_b^\alpha \partial_\mu e_\alpha^a. \tag{9.21}$$

We can also write the Ricci tensor as:

$$R_{\mu\nu} = e_\nu^a e_b^\sigma R_{\mu\sigma a}^b. \tag{9.22}$$

Hence we can write the action in terms of the triads and spin connection instead of the metric. Einstein Hilbert action is given as:

$$S_{EH} = \frac{1}{16\pi G} \int d^3x \sqrt{-g} \left(R - \frac{2}{\ell^2} \right). \tag{9.23}$$

This when rewritten in terms of triad and the spin connection, we get:

$$S_{EH} = \frac{1}{16\pi G} \int \left(e^a \wedge (2d\omega_a + \epsilon_{abc} w^b \wedge \omega^c) + \frac{1}{3\ell^2} \epsilon_{abc} e^a e^b e^c \right). \tag{9.24}$$

To obtain the gauge theoretic formulation, we define connection one forms, $A, \bar{A}$ through:

$$A = \omega + \frac{i}{\ell} e = \left(\omega_\mu^a + \frac{i}{\ell} e_\mu^a \right) T_a \, dx^\mu$$

$$\bar{A} = \omega - \frac{i}{\ell} e = \left(\omega_\mu^a - \frac{i}{\ell} e_\mu^a \right) T_a \, dx^\mu$$

where T_a are generators of $SL(2, R)$ (or $SU(2)$ in the case of Euclidean gravity). Using these one forms the gravity action becomes the Chern Simons action for A and $\bar{A}$.

Einstein Hilbert Gravity action is $S_{EH} = \frac{k}{2\pi} \left(S[A] - S[\bar{A}] \right)$ and

$$S[A] = \int A \wedge dA - \frac{2}{3} A \wedge A \wedge A \tag{9.25}$$

where $k = \frac{\ell}{8G}$. Euclidean action changes k to $-\frac{\ell}{8G}$ which is obtained by the continuation of $G \to -G$.

9.2.1 BTZ Blackhole

As mentioned earlier coset spaces obtained through identifications are still locally solutions of Einsteins equations viz., constant negative curvature geometries. But identification introduces horizons and hence correspond to blackhole solutions specified by two parameters, J and M, angular momentum and the mass. We will show that BTZ black hole in Euclidean continuation has the topology of a solid torus [4]. Since we are interested in the thermodynamic properties of the blackhole computed through path integrals, we will consider Euclidean blackhole [5] The metric for the Euclidean BTZ black hole in the usual Schwarzschild-like coordinates is obtained by the identifications mentioned earlier is [5–7]:

$$ds^2 = N^2 d\tau^2 + N^{-2} dr^2 + r^2 (d\phi + N^\phi d\tau)^2.$$

Here τ is the Euclidean time coordinate related to usual time t through $t = i\tau$ and

$$N = \left(-M + \frac{r^2}{l^2} r + - \frac{J^2}{4r^2}\right)^{\frac{1}{2}}, \quad N_\phi = -\frac{J}{2r^2}.$$

The inner and the outer horizons (obtained as solutions of $N = 0$) of the Lorentzian black hole solution get mapped in the Euclidean continuation to ir_- and r_+ respectively, where

$$r_\pm^2 = \frac{Ml^2}{2}\left[1 \pm \left(1 - \frac{J^2}{M^2 l^2}\right)^{1/2}\right]. \tag{9.26}$$

M and J, are as specified earlier, the mass and angular momentum of the black hole respectively. As shown by Carlip and Teitelboim [4], after a coordinate transformation,

$$x = \left(\frac{r^2 - r_+^2}{r^2 - r_-^2}\right)^{1/2} \cos\left(\frac{r_+}{l^2}\tau + \frac{|r_-|}{l}\phi\right) \exp\left\{\frac{r_+}{l}\phi - \frac{|r_-|}{l^2}\tau\right\},$$

$$y = \left(\frac{r^2 - r_+^2}{r^2 - r_-^2}\right)^{1/2} \sin\left(\frac{r_+}{l^2}\tau + \frac{|r_-|}{l}\phi\right) \exp\left\{\frac{r_+}{l}\phi - \frac{|r_-|}{l^2}\tau\right\},$$

$$z = \left(\frac{r_+^2 - r_-^2}{r^2 - r_-^2}\right)^{1/2} \exp\left\{\frac{r_+}{l}\phi - \frac{|r_-|}{l^2}\tau\right\}.$$

The black hole metric is the metric for hyperbolic three-space $\mathcal{H}_3$

$$ds^2 = \frac{l^2}{z^2}(dx^2 + dy^2 + dz^2), \quad z > 0,$$

9.2 Chern Simons Formulation of 3D Gravity

We transform this to spherical coordinates

$$x = R\cos\theta\cos\chi, \quad y = R\sin\theta\cos\chi, \quad z = R\sin\chi, \quad (9.27)$$

and the metric is

$$ds^2 = \frac{l^2}{R^2\sin^2\chi}\left[dR^2 + R^2 d\chi^2 + R^2\cos^2\chi\, d\theta^2\right], \quad (9.28)$$

Now we make global identifications to account for the periodicity of the ϕ coordinate in (9.26). These are easily seen to be:

$$(\ln R,\, \theta,\, \chi) \sim \left(\ln R + \frac{2\pi r_+}{l},\, \theta + \frac{2\pi|r_-|}{l},\, \chi\right).$$

Now we need to provide the Chern Simons connections in terms of the triads and spin connections. The connections A^a, $a = 1, 2, 3$ corresponding to the metric (9.28) may be written as:

$$A^1 = -\csc\chi\left(d\theta - i\frac{dR}{R}\right),$$

$$A^2 = i\csc\chi\, d\chi,$$

$$A^3 = i\cot\chi\left(d\theta - i\frac{dR}{R}\right).$$

The Chern-Simons formulation of gravity was used to describe the BTZ black hole first in [7], where for the Lorentzian black hole, the corresponding gauge fields were provided.

9.2.2 BTZ Blackhole and Statistical Mechanics

This formulation of 3D gravity with negative cosmological constant and BTZ blackhole solution through topological Chern Simons gauge theory and gauge fields with the corresponding gauge group provides a novel mechanism to compute the blackhole entropy. This will provide not only the leading term proportional to the boundary circumference but logarithmic corrections too.

Now we will proceed to the details of computations following [5]. We have already pointed out that the Euclidean BTZ blackhole has the topology of solid torus with a torus boundary. For entropy we evaluate the partition function of Chern Simons action on such a manifold with boundary. We have also expressed in the Chap. 7 that following Witten, the partition function on manifold with a boundary is described by WZW theory on the two dimensional boundary.

Fig. 9.1 Euclidean blackhole as solid torus

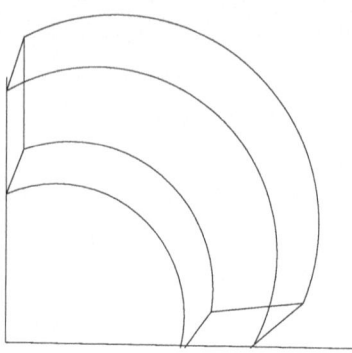

In order to compute the black hole partition function, we first evaluate the Chern-Simons path integral on the solid torus shown in Fig. 9.1. This path integral has been discussed in [8], [9], and many others.

We can use the gauge freedom and set the connection to a constant value on the torus boundary using gauge transformations. We introduce new coordinates to facilitate the computation instead of x and y with unit period. It is defined through $z = (x + \tau y)$ such that

$$\int_A dz = 1, \quad \int_B dz = \tau \qquad (9.29)$$

where A is the contractible cycle and B the non-contractible cycle of the solid torus. Also $\tau = \tau_1 + i\tau_2$ is the modular parameter of the boundary torus. We should remember a torus has two non contractible cycles. This is seen from $\pi_1(T_2) = \mathcal{Z} \oplus \mathcal{Z}$. But solid torus has only one. Then, the connection can be written as [9]:

$$A = \left(\frac{-i\pi \tilde{u}}{\tau_2} d\bar{z} + \frac{i\pi u}{\tau_2} dz \right) T_3 \qquad (9.30)$$

where u and $\tilde{u}$ are canonically conjugate fields and obey the canonical commutation relation following the canonical quantisation explained in Chap. 7 (Eq. 7.22):

$$[\tilde{u}, u] = \frac{2\tau_2}{\pi(k+2)}. \qquad (9.31)$$

They can be related to the black hole parameters by computing the holonomies of A around the contractible and non-contractible cycles of the solid torus.

9.2 Chern Simons Formulation of 3D Gravity

These holonomies have been computed in [4] for the general case of a rotating BTZ black hole solution with a conical singularity (Θ) at the horizon such that the identifications (9.29) characterizing the black hole now generalize to

$$(\ln R, \theta, \chi) \sim (\ln R, \theta + \Theta, \chi)$$

$$(\ln R, \theta, \chi) \sim \left(\ln R + \frac{2\pi r_+}{l}, \theta + \frac{2\pi |r_-|}{l}, \chi\right).$$

The former identification corresponds to the A cycle and the latter to the B cycle. Then the trace of the holonomies around the contractible cycle A and non-contractible cycle B are:

$$Tr(H_A) = 2\cosh(i\Theta), \qquad Tr(H_B) = 2\cosh\left(\frac{2\pi}{l}(r_+ + i|r_-|)\right). \tag{9.32}$$

For the classical black hole solution, $\Theta = 2\pi$. From (9.30),

$$A_z = \frac{-i\pi}{\tau_2}\tilde{u}, \qquad A_{\bar{z}} = \frac{i\pi}{\tau_2}u \tag{9.33}$$

where

$$u = \frac{-i}{2\pi}\left(-i\Theta\tau + \frac{2\pi(r_+ + i|r_-|)}{l}\right),$$

$$\tilde{u} = \frac{-i}{2\pi}\left(-i\Theta\bar{\tau} + \frac{2\pi(r_+ + i|r_-|)}{l}\right).$$

We need to exercise caution at this point. Here $\tilde{u}$ is the canonical conjugate, but not the complex conjugate of u. This is so because A is a complex $SU(2)$ connection.

Given the above machinery we can write the Chern-Simons path integral on a solid torus with a boundary modular parameter τ. For a fixed boundary value of the connection, i.e. a fixed value of u, this path integral is formally equivalent to a state $\psi_0(u, \tau)$ with no Wilson lines in the solid torus.

The states corresponding to having closed Wilson lines (along the non-contractible cycle) carrying spin $j/2$ ($j \leq k$) representations in the solid torus are given by [8], [9]:

$$\psi_j(u, \tau) = \exp\left\{\frac{\pi k}{4\tau_2}u^2\right\}\chi_j(u, \tau), \tag{9.34}$$

where χ_j are the Weyl-Kac characters [10] for affine SU(2). The Weyl-Kac characters can be expressed in terms of the well-known theta functions [10] as

$$\chi_j(u,\tau) = \frac{\Theta^{(k+2)}_{j+1}(u,\tau,0) - \Theta^{(k+2)}_{-j-1}(u,\tau,0)}{\Theta^{2}_{1}(u,\tau,0) - \Theta^{2}_{-1}(u,\tau,0)} \tag{9.35}$$

where theta functions are defined by:

$$\Theta^k_\mu(u,\tau,z) = \exp(-2\pi i k z) \sum_{n \in \mathbb{Z}} \exp 2\pi i k \left[\left(n + \frac{\mu}{2k}\right)^2 \tau \right.$$
$$\left. + \left(n + \frac{\mu}{2k}\right) u \right]. \tag{9.36}$$

The black hole partition function is to be constructed from the boundary state $\psi_0(u,\tau)$. To do that, we note the following:

1. Choose the microcanonical ensemble which is appropriate here. This corresponds in our picture, to keeping the holonomy around the non-contractible cycle B fixed [11]. The holonomy around the contractible cycle A is Θ, which has a value 2π for the classical solution. But it is *not* held fixed any more, and we sum over contributions to the partition function from all values of Θ. This amounts to we start with the value of u for the classical solution, $\Theta = 2\pi$ in (9.32), and translation of u of the form

$$u \to u + \alpha \tau \tag{9.37}$$

where α is an arbitrary number. This is obviously implemented by a translation operator:

$$T = \exp\left(\alpha \tau \frac{\partial}{\partial u}\right) \tag{9.38}$$

But this translation operator T is not gauge invariant. The only gauge-invariant way of implementing these translations is through Verlinde operators of the form

$$W_j = \sum_{n \in \Lambda_j} \exp\left(\frac{-n\pi \bar\tau u}{\tau_2} + \frac{n\tau}{k+2} \frac{\partial}{\partial u}\right) \tag{9.39}$$

where $\Lambda_j = -j, -j+2, \ldots, j-2, j$. This means that not all possible translations of u are allowed. Gauge invariance, allows only a subset of translations.

$$u \to u + \frac{n\tau}{k+2} \tag{9.40}$$

9.2 Chern Simons Formulation of 3D Gravity

where $n \in \mathcal{Z} \leq k$. Thus, the only allowed values of Θ are $2\pi(1+\frac{n}{k+2})$. Verlinde operator W_j [12] acting on a state without any Wilson line in the solid torus inserts a Wilson line of spin $j/2$ around the non-contractible cycle. Thus, taking into account all states with different translated values of u implies that we have to take into account all the states in the boundary with the insertion of such Wilson lines. These are the states $\psi_j(u, \tau)$ (9.34).

2. The final partition function, is obtained by integrating over the modular parameter, i.e. over all inequivalent tori with the same holonomy around the non-contractible cycle. The integrand, which is a function of u and τ, must be the square of the partition function of a gauged $SU(2)_k$ Wess-Zumino model corresponding to the two $SU(2)$ Chern-Simons theories. It also must be modular invariant. Modular invariance corresponds to large diffeomorphisms of the torus. That is, the partition function must remain invariant under a modular transformation.

The partition function is then of the form

$$Z = \int d\mu(\tau, \bar{\tau}) \left| \sum_{j=0}^{k} a_j(\tau) \psi_j(u, \tau) \right|^2 \tag{9.41}$$

where $d\mu(\tau, \bar{\tau})$ is the modular invariant measure, and the integration is over a fundamental domain in the τ plane. Coefficients $a_j(\tau)$ must be chosen such that the integrand is modular invariant. Under an $\mathcal{S}$ modular transformation, $\tau \to -1/\tau$ and $u \to u/\tau$, the $SU(2)_k$ characters transform as

$$\chi_j(u, \tau) \to \exp\left(-2\pi i k \frac{u^2}{4\tau}\right) \chi_j\left(\frac{u}{\tau}, \frac{-1}{\tau}\right)$$

$$= \exp\left(-2\pi i k \frac{u^2}{4\tau}\right) \sum_l S_{jl}\, \chi_l(u, \tau) \tag{9.42}$$

where matrix S_{jl} given by

$$S_{jl} = \sqrt{\frac{2}{k+2}} \sin\left[\frac{\pi(j+1)(l+1)}{k+2}\right], \qquad 0 \leq j, l \leq k \tag{9.43}$$

From the orthogonality of S matrix, we have the identity:

$$\sum_j S_{lj}\, S_{jp} = \delta_{lp}. \tag{9.44}$$

We are interested in the transformation property of the state $\psi_j(u, \tau)$ under an $\mathcal{S}$ modular transformation. The prefactor in (9.34) transforms into itself under such a transformation apart from an extra piece that exactly cancels the prefactor in (9.42). Thus, under an $\mathcal{S}$ transformation ($\tau \to -1/\tau$),

$$\psi_j(u, \tau) \to \sum_l S_{jl}\, \psi_l(u, \tau). \tag{9.45}$$

Under a $\mathcal{T}$ modular transformation ($\tau \to \tau + 1$), $\psi_j(u, \tau)$ picks up a phase,

$$\psi_j(u, \tau) \to \exp(2\pi i m_j)\, \psi_j(u, \tau) \tag{9.46}$$

where $m_j = \frac{(j+1)^2}{2(k+2)} - \frac{1}{4}$. For the integrand in (9.41) to be modular invariant, the coefficient $a_j(\tau)$ must transform under the $\mathcal{S}$ transformation as $a_j(\tau) \to \sum_p a_p(\tau) S_{pj}$ and under the $\mathcal{T}$ transformation as $a_j(\tau) \to \exp(-2\pi i m_j) a_j(\tau)$ so as to cancel the extra terms. Further, since the integrand must correspond to the square of the partition function of a gauged $SU(2)_k$ Wess-Zumino model, the coefficients $a_j(\tau)$ are just the complex conjugate of $SU(2)_k$ characters corresponding to $u = 0$, i.e., $(\psi_j(0, \tau))^*$. The black hole partition function therefore is

$$Z_{bh} = \int d\mu(\tau, \bar{\tau}) \left| \sum_{j=0}^{k} (\psi_j(0, \tau))^* \psi_j(u, \tau) \right|^2 \tag{9.47}$$

Finally the modular invariant measure is

$$d\mu(\tau, \bar{\tau}) = \frac{d\tau d\bar{\tau}}{\tau_2^2} \tag{9.48}$$

Thus we have obtained an exact expression Eq. (9.47) for the partition function of Euclidean black hole. We can check whether this corresponds to the semiclassical result.

We compare the above with the semi-classical entropy of black hole, by evaluating the expression (9.47) for large horizon radius r_+ by the saddle-point approximation.

Substituting from (9.34), (9.35) and (9.36), the saddle point of the integrand occurs when τ_2 is proportional to r_+ and therefore large. But for τ_2 large, the character χ_j is

$$\chi_j(\tau, u) \sim \exp\left[\frac{\pi i \left(\frac{(j+1)^2}{k+2} - \frac{1}{2}\right)}{2} \tau\right] \frac{\sin \pi(j+1)u}{\sin \pi u} \tag{9.49}$$

We now use in (9.47) the form of the character for large τ_2 from (9.49). In the expression for u in (9.32), we replace Θ by its classical value 2π. The computation

9.2 Chern Simons Formulation of 3D Gravity

has been done with positive coupling constant k and at the end, we must perform an analytic continuation to the Lorentzian black hole, by taking $G \to -G$. It can be checked that after the analytic continuation, it is the spin $j = 0$ in the sum over characters in (9.47) that dominates the partition function.

We obtain the leading behaviour of the partition function (9.47) for large r_+ (and when $|r_-| \ll r_+$) by first performing the integration over τ_1 in this regime. The τ_2 integration is done by the method of steepest descent. The saddle-point is at $\tau_2 = r_+/l$. Expanding around the saddle-point, by writing $\tau_2 = r_+/l + x$ and then integrating over x, we obtain

$$Z_{bh} = \frac{l^2}{r_+^2} \exp\left(\frac{-2\pi k r_+}{l}\right) \int dx \, \exp\left[-\frac{\pi k l}{2r_+} x^2\right] \tag{9.50}$$

upto a multiplicative constant. The logarithm of this expression yields the black hole entropy for large horizon length r_+:

$$S = \frac{2\pi r_+}{4G} - \frac{3}{2} \log\left(\frac{2\pi r_+}{4G}\right) + \ldots \tag{9.51}$$

The leading contribution to the black hole entropy which is proportional to the 'Area' is the familiar Bekenstein-Hawking term. The next-order correction to the semi-classical entropy is the logarithm of the black hole area $2\pi r_+$. The coefficient $-3/2$ of this correction is in agreement with that of the logarithmic correction of semi-classical entropy of four dimensional Schwarzschild black hole first observed in Ref. [13] in the quantum geometry formulation of gravity. The semi-classical Bekenstein-Hawking entropy for Euclidean BTZ black hole was previously studied in the path integral formulation in Ref. [6], but the logarithmic correction was not seen there. As described above, the right logarithmic correction is obtained by considering the correct modular invariant measure while integrating over all inequivalent tori (as the holonomy around the non-contractible cycle is held fixed).

The calculation presented here should be contrasted with an earlier calculation of partition function of a BTZ black hole coupled to a scalar field [14]. This is a perturbative one-loop calculation which incorporates a specific type of fluctuation, namely a scalar field. For small r_+, this leads to a different coefficient of the logarithmic correction in the entropy. On the other hand, calculation presented here following [5] is exact; it includes all possible quantum gravity fluctuations. It is therefore not surprising that the results differ.

It is interesting that AdS gas partition function can also be obtained from suitable limit of Eq. (9.47). As is well known [14], the action for the AdS gas can be obtained from that of the BTZ black hole by a transformation. For the case of a non-rotating black hole, this transformation has the form $r_+/l \to l/r_+$. With this change, the AdS gas partition function is

$$Z_{AdS}[r_+] = \int d\mu(\tau, \bar{\tau}) \left| \sum_{j=0}^{k} (\psi_j(0, \tau))^* \, \psi_j(u', \tau) \right|^2 \qquad (9.52)$$

where $u' = \frac{-i}{2\pi}\left(-i2\pi\tau + \frac{2\pi l}{r_+}\right)$.

The AdS gas partition function can again be evaluated by the saddle-point method. Small r_+ leads to a saddle-point with τ_2 large. In this limit of small r_+ (i.e., low temperature), the partition function is

$$Z_{AdS}[r_+] = \left(\frac{r_+}{l}\right)^{\frac{3}{2}} \exp\left(\frac{2\pi l^2}{4r_+ G}\right). \qquad (9.53)$$

This, at the leading order, agrees with the corresponding expression obtained in Ref. [14].

9.2.3 dS_3 Gravity and Chern Simons Formulation

When the cosmological constant is positive, the 3D spacetime becomes a de Sitter space dS_3. This is unlike the case of negative cosmological constant leading to AdS_3 space-time. dS_3 space has a cosmological horizon. Like AdS_3 and BTZ blackhole, we can study gravity in dS_3 also in Chern Simon's formulation [15]. We now explain its implications for a cosmological horizon. In this section, we will also provide the partition function for quantum gravity on dS_3 in an Euclidean path integral approach.

Global $(2+1)D$ de Sitter spacetime is described by the metric

$$ds^2 = -l^2 d\tau^2 + l^2 \cosh^2 \tau d\Omega^2 \qquad (9.54)$$

where $d\Omega^2$ stand for the S^2 metric. Equal time sections of this metric are two-spheres, and there are no globally timelike Killing vectors. However, there does exist a timelike Killing vector in certain patches of this spacetime.

Figure 9.2 shows the Penrose diagram of global de Sitter space with these patches—II and IV. These regions are causally disconnected and the timelike Killing vector flows in opposite directions in these two patches. Each of these patches is bounded by the cosmological horizon, and described by the metric

$$ds^2 = -N^2 dt^2 + N^{-2} dr^2 + r^2 d\phi^2 \qquad (9.55)$$

9.2 Chern Simons Formulation of 3D Gravity

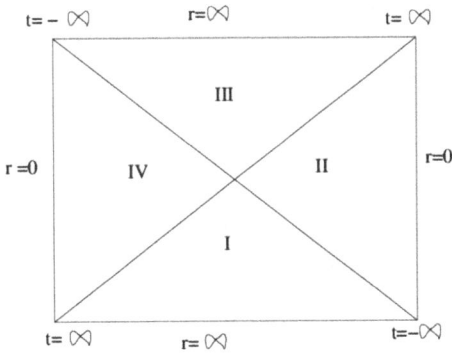

Fig. 9.2 Penrose diagram for dS_3

where

$$N^2 = (1 - \frac{r^2}{l^2}), \tag{9.56}$$

and $0 \leq r \leq l$, ϕ is an angular coordinate with period 2π. Since this metric is static, the patches II and IV are referred to as static patches. The cosmological horizon in these coordinates is therefore at $r = l$. Constant t surfaces are discs $\mathbf{D}_2$, and the topology of the patch is $D_2 \otimes \mathbb{R}$.

The gravity action I_{grav} in three dimensions written in a first-order formalism (using triads e and spin connection ω) is the difference of two Chern-Simons actions as described earlier. For Lorentzian gravity with a positive cosmological constant,

$$I_{grav} = I_{CS}[A] - I_{CS}[\bar{A}], \tag{9.57}$$

where

$$A = \left(\omega^a + \frac{i}{l} e^a\right) T_a, \quad \bar{A} = \left(\omega^a - \frac{i}{l} e^a\right) T_a \tag{9.58}$$

are SL(2, **C**) gauge fields (with $T_a = -i\sigma_a/2$). Here, the positive cosmological constant $\Lambda = (1/l^2)$. The Chern-Simons action $I_{CS}[A]$ is

$$I_{CS} = \frac{k}{4\pi} \int_M \mathrm{Tr}\left(A \wedge dA + \frac{2}{3} A \wedge A \wedge A\right) \tag{9.59}$$

and the Chern-Simons coupling constant is $k = -l/4G$.

9.2.4 Euclidean de Sitter Space in 3D

The Euclidean gravity action is the difference of two $SU(2)$ Chern-Simons actions—where the connections corresponding to the two actions are real and given by $A = \left(\omega^a + \frac{1}{l} e^a\right) T_a$ and $B = \left(\omega^a - \frac{1}{l} e^a\right) T_a$. We therefore consider the Euclidean continuation of the metric on the static patch given by Eq. (9.55). This is obtained by taking $t_E = it$. The metric is

$$ds^2 = N^2 \, dt_E^2 + N^{-2} \, dr^2 + r^2 \, d\phi^2 \, . \tag{9.60}$$

The periodicity of the Euclidean time now changes the topology of each of the static patches II and IV from the Lorentzian $\mathbf{D}_2 \otimes \mathbb{R}$ to $\mathbf{D}_2 \otimes \mathbf{S}^1$, where the $\mathbf{S}^1$ direction is the compactified Euclidean time. This is nothing but a solid torus.

Thus we are interested in the Euclidean gravity partition function studied through two $SU(2)$ Chern-Simons theories on a solid torus. Corresponding to the metric (9.60), the connections for the two $SU(2)$ Chern-Simons theories are given by:

$$A^0 = -N(d\phi + \frac{1}{l} dt_E)$$

$$A^1 = \frac{1}{lN} dr$$

$$A^2 = -\frac{r}{l^2} dt_E + \frac{r}{l} d\phi \, . \tag{9.61}$$

$$B^0 = N(-d\phi + \frac{1}{l} dt_E)$$

$$B^1 = -\frac{1}{lN} dr$$

$$B^2 = -(\frac{r}{l^2} dt_E + \frac{r}{l} d\phi) \, . \tag{9.62}$$

The procedure is the same as in AdS_3. For details refer to [15] and the previous section. The partition function can be worked out to be:

$$Z_{dS} = \int_{-1/2}^{1/2} d\tau_1 \, 4\pi \, e^{\beta \, k/2l} \, \frac{1}{f(\tau_1)} \, K_1(-k/2 \, f(\tau_1)) \, , \tag{9.63}$$

where $f(\tau_1) = \sqrt{\frac{\beta^2}{l^2} - 4\pi^2 \tau_1^2}$, and K_1 is the Bessel function of imaginary argument. Using the approximation for the Bessel function with large argument

$$K_1(z) = \sqrt{\frac{\pi}{2z}} e^{-z} [1 + O\left(\frac{1}{z}\right) + \ldots] \tag{9.64}$$

with the replacement $\beta = 2\pi l$ for de Sitter space, we get, in the large k regime:

$$Z_{dS} = 4\sqrt{\pi} \, \frac{4G}{2\pi l} \, e^{2\pi l/4G} \, . \tag{9.65}$$

9.2 Chern Simons Formulation of 3D Gravity

The form of the partition function indicates that at leading order, it is of the form e^S, where $S = \frac{2\pi l}{4G}$ is the semi-classical entropy. Since this is the partition function in the canonical ensemble, we would have expected an additional term $e^{-i\beta E}$ where E is the energy of de Sitter space. The notion of energy in asymptotically de Sitter spaces needs to be defined carefully, due to the absence of a global timelike Killing vector. The energy E that emerges in our formalism is defined on the horizon, and not at asymptotic infinity. This indicates that energy E is zero for de Sitter space. Such a result coincides with the definition of energy as given by Abbott and Deser [16].

The entropy is therefore

$$S = \frac{2\pi l}{4G} - \log \frac{2\pi l}{4G} + \ldots\ldots \quad (9.66)$$

The leading term is the semi-classical Bekenstein-Hawking entropy that is proportional to the horizon "area". The second term is the leading correction that is logarithmic in area.

9.2.5 Logarithmic Correction to the Entropy

The computation of entropy for BTZ blackhole and de Sitter was done in the same (Chern-Simons) formulation using the connection of 3D gravity to topological Chern Simons theory. The leading term as expected in both the cases was Bekenstein Hawking entropy and proportional to the 'area'. The first subleading term is interesting logarithmic and the numerical coefficient was $-3/2$ for BTZ blackhole, whereas for the de Sitter case, it was -1. This puzzling situation could be understood easily if we follow the scales in the two cases. The black hole entropy was computed in the regime $r_+ \gg \ell$, where r_+ is the black hole horizon radius and ℓ is the AdS radius of curvature. Then, there was an integral over the modular parameter similar to (9.52). The saddle-point for τ_2, the imaginary part of the modular parameter occurred when $\tau_2 = r_+/\ell$. Thus this was the regime when τ_2 was large. Replacing r_+/l in the black hole partition function by the inverse namely ℓ/r_+, where now $r_+ \ll \ell$, the AdS gas partition function was obtained, with the coefficient of the correction being $+3/2$. This corresponds to a situation where the modular parameter $\tau_2 = r_+/\ell$ is small. What happens when $r_+ \sim \ell$, i.e. $\tau_2 \sim 1$? In fact, this is very similar to the de Sitter case, since the de Sitter horizon radius is exactly ℓ!. It can in fact be verified directly that the saddle-point is at $\tau_2 = 1$. Hence we see that the coefficient of the logarithmic correction is -1. Thus, the coefficient of the correction in a semiclassical description depends on the regime one is looking at. When, as in the above case, there are two independent length parameters l and r_+, only for $r_+ \gg l$ do we get the coefficient $-3/2$.

Summarising results for the BTZ black hole:

$$For\ r_+ \gg l \quad S = \frac{2\pi r_+}{4G} - \frac{3}{2} \log\left(\frac{2\pi r_+}{4G}\right) + \cdots$$

$$r_+ = l \quad S = \frac{2\pi r_+}{4G} - \log\left(\frac{2\pi r_+}{4G}\right) + \cdots$$

$$r_+ \ll l \quad S = \frac{2\pi l^2}{4r_+ G} + \frac{3}{2}\log\left(\frac{r_+}{l}\right) + \cdots \quad (9.67)$$

where the last expression in (9.67) for $r_+ \ll l$ is the entropy of the AdS gas.

This indicates a duality proposed between the Euclidean BTZ black hole and Lorentzian de Sitter spaces. The classical holonomy of the connection in the black hole case was related to the ratio r_+/ℓ. From the duality, this was also the holonomy of the connection in a de Sitter space with a point mass, the mass being related to the parameter r_+.

In particular, vacuum de Sitter space corresponds to $r_+ = \ell$. Although this duality is only at the level of *actions* (for a Lorentzian theory with positive Λ and a Euclidean one with negative Λ), BTZ black hole with horizon radius $r_+ = l$ and vacuum de Sitter space have the same entropy—atleast at the leading and sub-leading order!

9.2.6 Logarithmic Corrections and AdS/CFT

In this framework all the information about quantum gravity in the bulk is expected to be contained in the conformal field theory at past or future infinity. The CFT is described by considering all possible metric fluctuations keeping the asymptotic behaviour to be de Sitter space. It consists of two copies of Virasoro algebras, each with central charge $c = 3l/2G$. The eigenvalues of the Virasoro generators L_0 and $\bar{L}_0$ for de Sitter space are both equal to $l/8G$. Using the Rademacher expansion for modular forms, one can generalize the Cardy formula for growth of states in a CFT beyond the leading term. The entropy [17] obtained from a CFT with a given the central charge c and eigenvalue of the Virasoro generator $L_0 = N$, is given by

$$S_1 = S_0 - 3/2 \log S_0 + \log c + \ldots \quad (9.68)$$

where $S_0 = 2\pi\sqrt{\frac{c}{6}(N - \frac{c}{24})}$. This is the contribution from the Virasoro generator L_0. There is a similar contribution S_2 associated with the Virasoro generator $\bar{L}_0$, given by replacing N in the above by $\bar{N}$, the eigenvalue of $\bar{L}_0$.

Substituting $c = 3l/2G$ and $N = \bar{N} = l/8G$ in the above, we see that

$$S = S_1 + S_2 = \frac{2\pi l}{4G} - \log\frac{2\pi l}{4G} + \cdots \quad (9.69)$$

with the same coefficient -1 for the logarithmic correction as that obtained from the gravity partition function (9.65) in (9.66). Here, the contribution from each of S_1 and S_2 to the logarithmic correction was $-1/2 \log \frac{2\pi l}{4G}$.

Thus, the quantum gravity calculation of de Sitter entropy and the entropy computation from the asymptotic CFT agree even in the sub-leading correction to the Bekenstein-Hawking term.

Exercises

9.1. Show ζ given in Eq. (9.15) is a Killing vector.

9.2. Show the ζ leads to the values of Casimir invariants given in Eqs. (9.16, 9.17).

9.3. Show the Einstein Hibert action in terms of triads and spin connection is Eq. (9.24)

9.4. Show the connections corresponding to the BTZ metric in Eq. (9.28) can be rewritten as given in Eqs. (9.30, 9.33).

9.5. Prove Eq. (9.63).

References

1. F. Larsen, Lectures on Kerr/CFT, ICTP Lectures (2010)
2. A. Achucarro, P. Townsend, Phys. Lett. B **180**, 89 (1986), S. Deser, R. Jackiw, Ann. Phys. 153, 405 (1984)
3. M. Banados, C. Teitelboim, J. Zanelli, Phys. Rev. Lett. **69**, 1849 (1992); M. Banados, M. Henneaux, C. Teitelboim, J. Zanelli, Phys. Rev. **D48**, 1506 (1993)
4. S. Carlip, C. Teitelboim, Phys. Rev. **D51**, 622 (1995)
5. T.R. Govindarajan, R.K. Kaul, V. Suneeta, Class. Quant. Grav. **18**, 2877 (2001)
6. S. Carlip, Class. Quant. Grav. **16**, 3327 (1999)
7. D. Cangemi, M. Leblanc, R. Mann, Phys. Rev. **D48**, 3606 (1993)
8. S. Elitzur, G. Moore, A. Schwimmer, N. Seiberg, Nucl. Phys. B **326**, 108 (1989)
9. J.M. Isidro, J.M.F. Labastida, A.V. Ramallo, Nucl. Phys. B **398**, 187 (1993)
10. V. Kac, *Infinite-Dimensional Lie Algebras* (Cambridge University Press, 1994). D. Mumford, Tata Lectures on Theta, I, 2007, Springer
11. J.D. Brown, G.L. Comer, E.A. Martinez, J. Melmed, B.F. Whiting, J.W. York, Class. Quant. Grav. **7**, 1433 (1990)
12. E.P. Verlinde, Fusion rules and modular transformations in 2D conformal field theory. Nucl. Phys. **B300**, 360 (1988)
13. R.K. Kaul, P. Majumdar, Phys. Lett. **B439**, 267 (1998)
14. R. Mann, S. Solodukhin, Phys. Rev. **D55**, 3622 (1997)
15. T.R. Govindarajan, R.K. Kaul, V. Suneeta, Class. Quant. Grav. **19**, 4195 (2002)
16. L.F. Abbott, S. Deser, Nucl. Phys. **B195**, 76 (1982)
17. D. Birmingham, S. Sen, Phys. Rev. **D63**, 047501 (2000)

SPRINGER NATURE

GPSR Compliance

The European Union's (EU) General Product Safety Regulation (GPSR) is a set of rules that requires consumer products to be safe and our obligations to ensure this.

If you have any concerns about our products, you can contact us on ProductSafety@springernature.com

In case Publisher is established outside the EU, the EU authorized representative is:

Springer Nature Customer Service Center GmbH
Europaplatz 3
69115 Heidelberg, Germany

The manufacturer's authorised representative in the EU is Springer Nature Customer Service Centre GmbH, Europaplatz 3, 69115 Heidelberg, Germany. If you have any concerns regarding our products, please contact ProductSafety@springernature.com

Printed and bound by CPI Group (UK) Ltd, Croydon, CR0 4YY

26/03/2026

02078992-0005